Lab Manual

Understanding Food
Principles and Preparation

FIFTH EDITION

Amy C. Brown

Prepared by

Janelle M. Walter, Ph.D., R.D., C.F.C.S.
Baylor University

Karen Beathard, M.S., R.D., L.D.
Texas A & M University

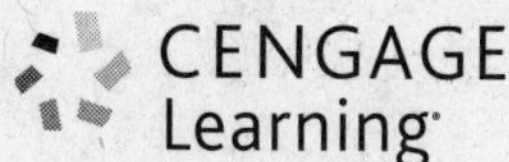

Australia • Brazil • Mexico • Singapore • United Kingdom • United States

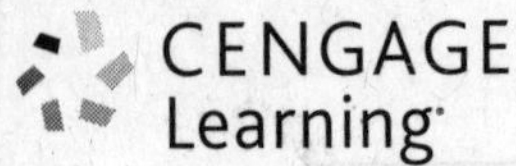

For product information and technology assistance, contact us at **Cengage Learning Customer & Sales Support, 1-800-354-9706**.

For permission to use material from this text or product, submit all requests online at **www.cengage.com/permissions**
Further permissions questions can be emailed to **permissionrequest@cengage.com**.

ISBN-13: 978-1-133-60716-8
ISBN-10: 1-133-60716-0

Cengage Learning
200 First Stamford Place, 4th Floor
Stamford, CT 06902
USA

Cengage Learning is a leading provider of customized learning solutions with office locations around the globe, including Singapore, the United Kingdom, Australia, Mexico, Brazil, and Japan. Locate your local office at: **www.cengage.com/global**.

Cengage Learning products are represented in Canada by Nelson Education, Ltd.

To learn more about Cengage Learning Solutions, visit **www.cengage.com**.

Purchase any of our products at your local college store or at our preferred online store **www.cengagebrain.com**.

Printed in the United States of America
5 6 7 8 9 10 11 20 19 18 17 16

Table of Contents

Preface

This edition of the lab manual is designed specifically to accompany Amy Brown's *Understanding Food: Principles and Preparation*, 5th edition. It is formatted with the student and course instructor in mind. Key terms that are applicable to each unit are identified and include a brief definition. Pre-lab questions have been prepared in order to encourage students to investigate or research or study the topic prior to the lab experience. Each lab includes relevant experiments with variations that enhance knowledge of functional properties of food. Tables to record data follow each experiment so that students can easily document observations from experiments. In addition, most units contain additional recipes for preparation practice. Lastly, each unit contains post-lab questions to help students evaluate the final products and discuss their observations.[1]

Appendices are included in the manual in order to enhance lab organization. Appendix A includes a description of various objective tests and forms that are required for assessment of products. Appendix B is designed to assist the course instructor in lab coordination. Units are arranged in numerical order and include a listing of ingredients and equipment required for each experiment and recipe. Appendix C is a glossary of key terms from each of the units. The alphabetical arrangement of key terms makes it easier for students to review culinary terms. Appendix D is a listing of figures (new to this edition) and tables by unit, and a recipe index is provided as Appendix E.

Our hope is that each student and course instructor will find this lab manual very user friendly and the knowledge that you gain will enhance your success in your professional career.

New to This Edition[2]

This edition retains the general format and organization of the 4th edition manual, but various enhancements and corrections have been made.

- Clarity of the instructions in several labs has been improved.
- A few of the existing exercises have been modified to include additional or more pertinent variations.
- Several new photos have been added: Figures 1-3 and 17-2 (texture analysis), 1-4 (a triangle test), 2-2 (measuring flour), 3-2 (a wok), 11-1 (preparing a gluten ball), 12-1 (technique for preparing a starch paste), 13-3 (measuring muffin height), and 20-1 and 20-2 (coffee and tea pots).
- The pre- and post-lab questions have been thoroughly revised to more closely mirror the information in the text and this manual, and chapter references are now provided.

---

[1] Note for instructors: An answer key for the lab questions is available on the instructor website and the Power Lecture (ISBN 9781133607182).

[2] Note for instructors: A detailed list of changes to the lab manual is available on the instructor website.

Food Lab Safety and Orientation

Objectives

1. To introduce the student to the purposes, policies, and procedures of the laboratory sessions.
2. To introduce the importance of safe food handling techniques.
3. To introduce the student to the equipment to be used in the laboratory.

Laboratory Purpose

The purpose of this laboratory is to provide "hands on" experience in food production. Following instructions and recipes (or formulas) for each lab will provide an opportunity to observe and explore the science of food production methods and quality outcomes.

Course Learning Outcomes

- Demonstrate safe preparation methods that preserve the quality and nutritive value of foods.
- Identify the functional and structural properties of various food groups including fats, starches, fruits and vegetables, eggs, dairy, meat, poultry, and fish.
- Identify appropriate terminology for food preparation and production and culinary techniques.
- Develop sensory evaluation skills and apply these to all aspects of food preparation and preservation.

General Laboratory Procedures

All food, equipment, and supplies required for lab will be available in a location identified by the instructor. In addition, the instructor will provide specific lab policies and procedures related to the handling of inventory that must be followed.

- It is critically important that the assigned lab is read prior to coming to lab.
- Each student should be familiar with the products they will be preparing in lab and the quantity of ingredients needed.
- Efficient practices should be followed in obtaining and handling product. For example, do not take more ingredients than you need and minimize the number of trips made to the ingredient table.
- In addition, specifically follow recipe directions so ingredients are not wasted and proper observations can be made.

Safe food handling practices must be followed throughout the lab.

- Thorough hand washing will significantly reduce the risk of food contamination and foodborne illness. Students must wash their hands upon entry to the lab and between tasks.
- Food handler gloves should be worn when handling ready-to-eat (RTE) foods.
- Serving utensils are required when portioning food samples.
- Dry hands with paper towels. Use dish cloths for the dishes if you are not air drying.

Proper food handling practices must also be followed. Temperatures of potentially hazardous foods (PHF) should be monitored. Hot foods should be maintained hot, cold foods maintained cold, and the danger zone (40° F to 135° F) should be avoided. Potential for cross-contamination must be minimized. Raw

PHF should never come into contact with RTE foods. Proper procedures for washing and sanitization of equipment including scales, measuring utensils, serving plates, etc. must be followed.

Each student is responsible for the cleanliness and sanitization of the lab. Everyone should clean their respective work area before leaving the laboratory. A specific cleaning protocol will be provided by the laboratory instructor.

Each student is responsible for personal safety. Students should take precautions when handling hot food and equipment, knives, and glassware. Students must become familiar with the proper operation and cleaning of equipment. Equipment should be turned off when not in use. The instructor will identify the location of fire extinguishers and first aid equipment.

Students should not leave the laboratory room for any reason without checking with the laboratory instructor. Students must check out with the laboratory instructor at the end of lab.

Personal Hygiene and Dress

The instructor will provide specific dress requirements for this laboratory. Good hygienic standards including clean apparel, fingernails, and hair should be maintained by all students.

A hair covering such as a hair net or clean ball cap is recommended.

Closed-toe, rubber-sole shoes are recommended; sandals, flip flops, and high heels generate potential safety issues and are discouraged.

Unit 1 – Sensory Evaluation

Introduction

Study of the chemical and physical properties of food components and the interactions among these components during preparation and storage provides a basis for understanding factors that influence food quality. Measuring quality differences in foods is an essential part of such a study. A number of objective tests have been devised for this purpose (Figures 1-1, 1-2, and 1-3 show objective tests being performed), but there are still many aspects of food quality for which no appropriate objective tests are available. Instead, subjective tests must be used to assess differences in the quality of foods. The ultimate criterion used to evaluate food quality characteristics such as appearance, texture, taste, and aroma is the degree of appeal to human sensory organs.

Figure 1-1: Correct use of a Brix refractometer to measure the concentration of a syrup.

Figure 1-2: Penetrometer measurement of a muffin.

Figure 1-3: Texture analyzer measurement of a muffin.

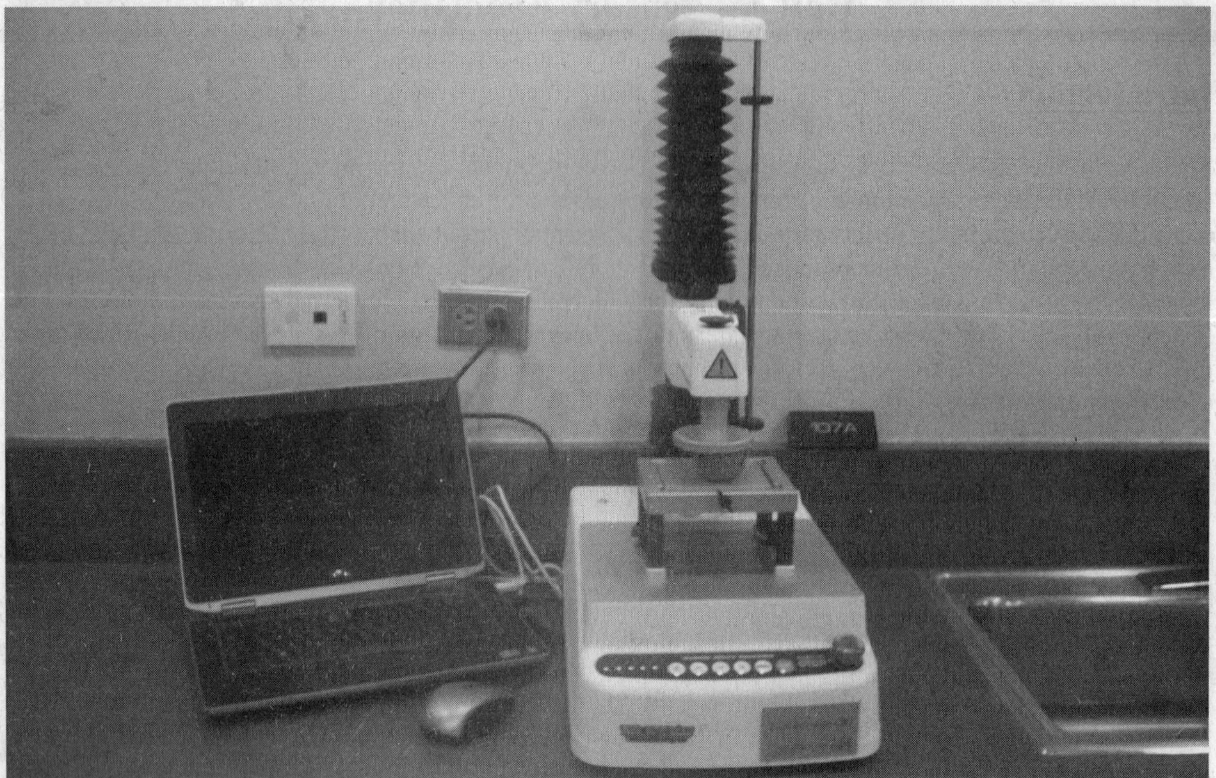

Sensory evaluation of foods has been applied to many situations in research and development, quality control, and food product marketing. Such applications include: new product development, product matching and improvement, process change, selection of a new source of supply, quality control, storage stability, product selection and training, and correlation of sensory with chemical and physical measurements.

Specific points to keep in mind during sensory evaluation studies include:

1. Each individual must work independently.
2. There must not be any discussion while foods are being evaluated.
3. Class discussion will be led by the lab instructor following individual evaluations.
4. The success of the scoring procedure depends on the degree to which the student can put himself/herself into the position of an impartial judge.
5. To avoid bias, samples are generally coded with 3-digit random code numbers. Samples should be representative of the product being tested.
6. If he or she is careless, the scorer may miss differences that should have been detected; on the other hand, if he or she is hypercritical, the scorer may record deviations that in reality do not exist.

A Guide for Food Product Evaluation

Quality is commonly thought of as degree of excellence. Food quality is a complex concept that includes appearance, flavor, texture, and nutrient content. In the food laboratory, *food quality* refers to the sensory characteristics of the food, which is evaluated subjectively by the human senses. The specific sensory characteristics that are evaluated include appearance, flavor, and texture. Taste and odor are components of flavor.

Appearance includes shape, size, color, and the condition of the outside surface of the food. For example, is it transparent, opaque, dull, glossy, and/or free of defects? In baked products, appearance also may include the color of the interior crumb.

Flavor is the total sensory impression when food is eaten. Flavor is a combination of taste and aroma as well as mouthfeel sensations.

Taste is induced by gustatory sensations. Taste buds are stimulated by substances in solution and are believed to respond to four primary tastes—bitter, salty, sour, and sweet. It is unclear whether umami is a fifth basic taste or a flavor enhancer. All other tastes are believed to be a combination of these. Astringency is a taste-related phenomenon that causes the mouth to pucker and is associated with sourness. The sense of taste is weak compared to the sense of smell. Taste is often more easily perceived in foods that are of fluid consistency and in liquids.

Aroma is the component of flavor that strongly influences acceptance or rejection of a food. The sense of smell is very keen. Only a few molecules of a volatile compound are necessary to produce an olfactory sensation. Aroma is modified by food temperature. Higher temperatures result in greater volatility of flavor compounds and therefore in greater flavor perception.

Texture is the response of the tactile senses to the physical characteristics of a food. Interpretation of these tactile sensations (touch, pressure, and movement) is what we generally refer to as texture. Our evaluation of a food's texture is influenced by both the structure of the food and by its resistance to forces applied by the teeth, tongue, roof of the mouth, knife, and fork—in other words, the food's tenderness. Consistency is also an aspect of texture and is defined by terms such *brittleness* or *chewiness*.

Mouthfeel is a specific sensation detected by the lining of the mouth, and it interacts with flavor perception. Examples of mouthfeel sensations include pain (jalapeno peppers, horseradish), temperature (hot, frozen), astringency (pickles, lemonade), and tactile stimulation (carbonated vs. flat beverage).

Descriptive Terms

The following adjectives are terms that may be used in sensory evaluation. This list of terms is by no means exhaustive. The terms merely serve as examples.

Appearance: symmetrical, asymmetrical, level, sunken, rounded, pebbled, sticky, greasy, shiny, dry, pale, golden brown, light brown, burnt, smooth, rough, puffy, transparent, dark, creamy, curdled, dull, fine, grainy, moist, sticky, opaque, glossy, clear, any color or shape.

Flavor: sweet, bitter, sour, salty, astringent, spicy, soapy, floury, flat, eggy, rancid, pasty, bland, flowery, fruity, sharp, burnt, minty, pungent, putrid, musky, puckery, hot, cold, metallic, burning (peppery), cool (minty), fishy, nutty, yeasty, stale, watery, luscious.

Texture: crisp, velvety, smooth, rough, hard, firm, thick, thin, viscous, springy, gritty, gummy, adhesive, moist, tender, fibrous, chewy, curdled, lumpy, pasty, rubbery, tough, greasy, fibrous, crunchy, smooth, creamy, gelatinized, stringy, flaky, crusty, limp, mealy, mushy.

Mouthfeel: crisp, sticky, slimy, gritty, slick, crunchy, smooth.

Objectives of Sensory Evaluation

When evaluating a food product, a natural reaction might be to say simply "I don't like it" or "it's really good." Statements such as these represent a summation of instantaneous evaluation of the product's appearance, flavor, and texture. In addition, an evaluation is undoubtedly influenced by the cultural, emotional, and psychological biases that play a role in determining individual food preferences. Because sensory evaluation of food products is subjective, it can subconsciously be influenced by the evaluator's personal biases. The desired objectives of sensory evaluation in the food laboratory are to set individual preferences aside, consciously make an evaluation based on only the relevant sensory characteristics of appearance, flavor, and texture, and communicate your evaluation to others by using appropriate descriptive terms.

Key Terms

analytical tests – subjective tests that are used to detect differences. Examples include discriminative tests and descriptive tests.

affective tests – subjective tests that evaluate individual preferences. Examples include hedonic and personal preference tests.

chemethesis – phenomenon in which certain foods that are not physically hot or cold give the impression of being "hot" or "cold" when placed on the tongue. An example is hot peppers.

threshold concentration – concentration required to elicit a taste response.

Pre-Lab Questions

Textbook reference: Chapters 1 and 2

1. Define sensory evaluation.

2. What senses are used to assess food quality?

3. What factors impact the taste of food?

4. Compare subjective and objective testing methods that may be used in research.

5. What is taste fatigue and how can it be avoided?

6. What type of taste panels can be used? How would selection of panelists differ among them?

7. What are the environmental guidelines for food evaluation?

Lab Procedures

A. Evaluation of Food Products Using Descriptive Terms

Descriptive tests are analytical tests that are used to detail specific flavors or textures of a food or beverage.

Objectives
1. To become familiar with descriptive terms used in sensory evaluation of foods.
2. To evaluate the appearance, consistency, flavor, aroma, and composition of various food products by using the human senses.

Basic Procedure to Evaluate Products Using Descriptive Terms

Ingredients
Assortment of food products (chosen by instructor)
2-oz. sample cups

Procedure
1. Place bite-size samples of selected products into 2-oz. sample cups. Prepare enough cups for each participant to sample each product.
2. Evaluate the appearance, aroma, flavor, texture, and consistency of each product by using the descriptive terms provided above or other appropriate terms. Record observations in Table A-1.

Table A-1 Evaluation of Food Products Using Descriptive Terms

Product	Appearance	Aroma	Flavor	Texture	Consistency

B. Paired Comparison Test

The **paired comparison test** is a difference test. Two coded samples are presented simultaneously and panelists are asked to select the one that has more of a particular characteristic (sweet, sour, thick, etc.). The chance of selecting the correct sample is one out of two.

Objectives

1. To conduct a paired comparison test and participate as a sensory panelist.
2. To determine which sample possesses a greater intensity of the characteristic being evaluated.

Basic Procedure for the Paired Comparison Test

Ingredients

Two similar food or beverage products with assigned sample codes (chosen by instructor)
2-oz. sample cups

Procedure

1. Place 1-oz. samples of selected products into 2-oz. sample cups marked with the respective sample codes. Prepare enough cups for each participant to sample both products.
2. Each participant should taste both samples and determine which sample has the greater intensity of the characteristic being evaluated.
3. Record data and observations in Table B-1.

Table B-1 Paired Comparison Test

Sample Code	Intensity (Lesser or Greater)	Characteristic Evaluated

C. Triangle Test

The **triangle test** (Figure 1-4) is a difference test performed on three coded samples. Two samples are identical and a third is different. All samples are presented simultaneously. Panelists are asked to identify the odd sample. Samples must be similar in appearance or this test is not very meaningful. The chance of guessing the odd sample is one out of three.

Figure 1-4: The triangle test is a test of difference.

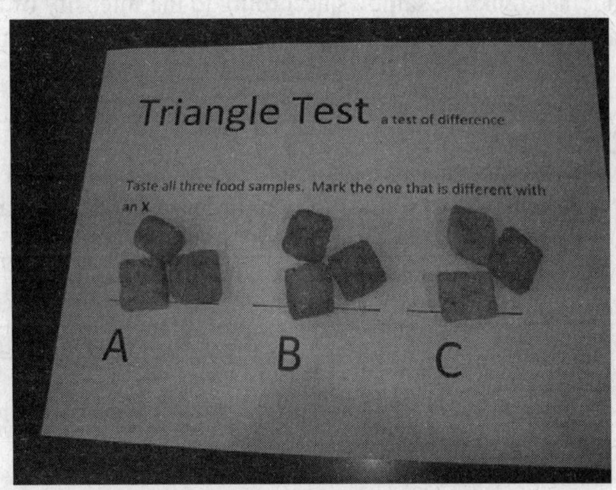

Objectives

1. To conduct a triangle test and participate as a sensory panelist.
2. To identify the odd sample of a series of three coded samples.

Basic Procedure for the Triangle Test

Ingredients

Two identical food or beverage products with assigned sample codes (chosen by instructor)
One similar food or beverage product with an assigned sample code (chosen by instructor)
2-oz. sample cups

Procedure

1. Place 1-oz. samples of selected products into 2-oz. sample cups marked with the respective sample codes. Prepare enough cups for each participant to sample all three products.
2. Each participant should taste each of the three coded samples and determine which sample differs from the other two.
3. Record data and observations on Table C-1.
4. Determine the total number of correct responses in the class and evaluate results as to the likelihood of selecting the odd sample by chance.

Table C-1 Triangle Test

Sample Code	Identify as Different or Same

D. Ranking Test

A **ranking test** is a difference test where more than two samples are simultaneously presented and the panelists rank the samples according to the intensity of the particular characteristic (flavor, odor, color, etc.). The sample with the greatest intensity is ranked number one.

Objectives

1. To conduct a ranking test and participate as a member of the sensory panel.
2. To rank a series of samples in order of intensity of a specific characteristic.
3. To rank a series of samples in order of preference.

Basic Procedure for the Ranking Test

Ingredients

3-5 similar food or beverage products with varying intensity of specific characteristics and an assigned sample code (chosen by instructor)
2-oz. sample cups

Procedure

1. Place 1-oz. samples of selected products into 2-oz. sample cups marked with the respective sample codes. Prepare enough cups for each participant to sample all products.
2. Each participant should taste each of the coded samples and rank them in descending order of intensity with the most intense sample ranked #1.
3. You may re-taste any of the samples while ranking for intensity of the specific characteristics; no ties are allowed in the ranking.
4. Record results in Table D-1.
5. Rank the samples again according to preference. The order may be the same as or different than the ranking for intensity.

Table D-1 Ranking Test

Ranking	Intensity	Preference
#1 – Most		
#2		
#3		
#4		
#5 – Least		

E. Identification of Primary Tastes

Objectives

1. To identify the four primary tastes.

Basic Procedure to Identify Primary Tastes

Ingredients

5 coded samples that include each of the following:

- Tonic water
- Sucrose solution
- Vinegar solution
- NaCl solution
- Duplicate solution (1 of the previous 4 with a different sample code)

2-oz. sample cups
Unsalted crackers
Water

Procedure

1. Place 1-oz. samples of the five coded solutions into 2-oz. sample cups marked with the respective sample codes. Prepare enough cups for each participant to sample all five solutions.
2. Each participant should taste each of the five coded samples and identify the primary taste of each.
3. Water and unsalted crackers should be used to rinse the mouth between samples.
4. Record the sample code identified as coinciding with each taste in Table E-1.
5. Record the correct code that coincides with each taste (as provided by the instructor) in Table E-1.

Table E-1 Identification of Primary Tastes

	Bitter	Sour	Salty	Sweet	Duplicate
Sample code					
Correct code					

F. Evaluation of Seasonings

Objectives

1. To study the characteristic flavor, color, and form of spices and herbs.
2. To study suggested uses and applications for spices and herbs.

F-a. Spiced Rice

Basic Recipe for Spiced Rice

Ingredients

1 c. rice, long-grain white (uncooked)
2 ¼ c. water
1 t. margarine
Assortment of spices (chosen by instructor)

Procedure

1. Heat water to boiling on high. Add rice. Cover with a lid and turn heat to medium-low. Cook 20 minutes.
2. Remove rice from heat and divide among 4 custard cups.
3. Stir ¼ t. margarine and a pinch of one of the selected spices into each custard cup of hot rice.
4. Display a small amount of each spice used in a labeled custard cup or on waxed paper, next to the cup of seasoned rice.
5. Evaluate seasoned rice for color, form, and palatability and record observations in Table F-1.

Table F-1 Evaluation of Spices in Rice

Spice	Color[1]	Form[2]	Palatability[3]	Suggested Uses

1. Color of the dry spice used; 2. Finely ground, coarsely ground, shredded, etc.; 3. Agreeableness of the taste based on personal preference

F-b. Tomato Bouillon

Basic Recipe for Tomato Bouillon

Ingredients

1 c. tomato juice
2 c. beef bouillon (made using bouillon cubes, liquid concentrate, or granules)
Assortment of herbs (chosen by instructor)

Procedure

1. Mix tomato juice and beef bouillon and heat to boiling.
2. Divide into small teapots or liquid measuring cups.
3. Stir $^1/_8$ t. of one of the selected herbs into each portion, cover, and steep for 3 minutes. Use just enough herbs to distinguish the characteristic flavor.
4. Serve hot.
5. Display some of each herb used in a labeled custard cup next to the cup of seasoned hot bouillon.
6. Evaluate tomato bouillon with herbs for color, form, and palatability and record observations in Table F-2.

Table F-2 Evaluation of Herbs in Tomato Bouillon

Herb	Color[1]	Form[2]	Palatability[3]	Suggested Use

1. Color of the herb displayed; 2. Finely ground, crushed leaves, broken leaves, etc.; 3. Agreeableness of the taste based on personal preference

F-c. Spiced Applesauce

Basic Recipe for Spiced Applesauce

Ingredients

1 16-oz. jar applesauce
Assortment of spices (chosen by instructor)

Procedure

1. Divide a large can or jar (16 oz.) of applesauce into small bowls or custard cups.
2. Stir $^{1}/_{8}$ t. of one of the selected ground spices into each portion of applesauce.
3. Use just enough spice to distinguish the characteristic flavor.
4. Display a small amount of each spice used in a labeled custard cup next to the seasoned applesauce.
5. Evaluate spiced applesauce for color, form, and palatability and record observations in Table F-3.

Table F-3 Evaluation of Spices in Applesauce

Spice	Color[1]	Form[2]	Palatability[3]	Suggested Uses

1. Color of the spice used; 2. Finely ground, coarsely ground, seeds, etc.; 3. Agreeableness of the taste based on personal preference

F-d. Seasoned Cream Cheese

Basic Recipe for Seasoned Cream Cheese

Ingredients

8 oz. cream cheese
1 ½ T. water
Assortment of herbs or seeds (chosen by instructor)
Unsalted crackers

Procedure

1. Cream 8 oz. of cream cheese with 1 ½ T. water to provide a fluffy, easy-to-spread mixture.
2. Divide the cheese into portions and add $^{1}/_{8}$ t. of the select herbs or seeds listed to each portion.
3. Add just enough of the herb or seed to distinguish the characteristic taste.
4. Serve the spread on unsalted crackers.
5. Display some of the herb or seed in a labeled custard cup next to the sample.
6. Evaluate seasoned cream cheese spread for color, form, and palatability and record observations in Table F-4.

Table F-4 Evaluation of Seasoned Cream Cheese				
Herb or Seed	Color[1]	Form[2]	Palatability[3]	Suggested Uses

1. Color of the herb or seed; 2. Whole seed, finely ground leaves, crushed leaves, whole leaves, etc.; 3. Agreeableness based on personal preference

Post-Lab Study Questions

Textbook reference: Chapters 1 and 2

Discussion Questions

1. What is the meaning of the saying "You eat with your eyes"?

2. Name one advantage and one possible disadvantage of using a paired comparison test in sensory evaluation testing in a lab.

Questions for Post-Lab Writing

3. How would you set up a triangle test to compare two types of vanilla yogurt (A and B)? What would the results tell you?

4. For what type of research is the triangle test best suited?

5. What are the primary tastes? If tonic water was used in lab, which two primary tastes were detectable?

6. What chemical compound(s) is/are thought to be responsible for the taste of bitter? Sour? Salty? Sweet?

7. What physical state must all substances be in before a taste can be detected?

8. What is a ranking test, and how is this test conducted?

9. What are your chances of guessing correctly in the paired comparison test? In the triangle test? How would this affect your interpretation of the results?

10. How would you evaluate your potential as a sensory analyst, based on results from this laboratory?

11. Were there any seasonings that you had not tried before? What was your evaluation of them?

12. Which type of evaluation was done with the seasonings?

Unit 2 – Food Preparation Basics: Measuring Techniques and Energy Transfer

Introduction

Quality is commonly thought of as degree of excellence. Food quality includes appearance, flavor, texture, wholesomeness, and nutrient content. Food quality is highly dependent on the preparation methods used in production. Knowledge of the functional and structural properties of food certainly enhances quality outcomes. However, appropriate application of basic food preparation methods including proper cooking techniques and equipment selection and accurate measurement and mixing procedures is essential to the production of high-quality food.

Key Terms

beat - to make a mixture smooth or to introduce air by whipping, using a rapid, regular motion.

blend – to mix two or more ingredients thoroughly.

calibrate – to confirm or compare a measurement with an established standard.

cream – to soften or blend one or more foods until creamy by beating with a spoon or electric mixer; e.g., sugar, shortening, and other ingredients are creamed together to incorporate air so that the resultant mixture increases appreciably in volume and is thoroughly blended.

dry-heat preparation methods – cooking techniques in which heat is transferred by air, radiation, fat, or metal. These include baking, roasting, broiling, grilling, barbequing, and frying.

meniscus – the curved upper portion at the surface of a liquid in a container that is caused by surface tension. It is concave if the liquid is attracted to the container walls, but it is convex if it is not.

moist-heat preparation methods – cooking techniques in which liquids such as water or a water-based liquid or steam are used to heat foods and enhance their appearance and palatability. These techniques include scalding, poaching, simmering, stewing, braising, boiling, parboiling, blanching, and steaming.

sift – to put dry ingredients through a sieve.

toast – to brown the surface of food by application of direct or dry heat.

Pre-Lab Questions

Textbook reference: Chapters 4 and 5

1. What is the difference between a dial meat thermometer and a dial candy or frying thermometer?

2. What measuring utensil provides the most accurate measurement of 1 c. milk? Of ½ c. granulated sugar?

3. How does the baking pan surface affect energy transfer and the final product when baking cookies?

4. What is the difference between 1 oz. and 1 fl. oz.?

5. Describe the differences between a conventional and convection oven.

6. Why is a "rest period" or "standing time" included in directions for microwave recipes?

Lab Procedures

A. Commonly Used Measurements

Objectives

1. To learn commonly used household measurements and abbreviations.
2. To learn metric equivalents of the commonly used measurements.
3. To learn how to convert temperatures on the Celsius scale (°C) to temperatures on the Fahrenheit scale (°F) and vice versa by using the following equations.

Procedure

1. Learn the lists of measurements and abbreviations in Table A-1.

Table A-1 Commonly Used Measurements and Abbreviations	
3 teaspoons (t, tsp.)	= 1 tablespoon (T, Tbsp.)
4 tablespoons (T, Tbsp.)	= ¼ cup (c.)
$5^1/_3$ tablespoons (T, Tbsp.)	= $^1/_3$ cup (c.)
16 tablespoons (T, Tbsp.)	= 1 cup (c.)
1 cup (c.)	= 8 fluid ounce (fl. oz.)
2 cups (c.)	= 1 pint (pt.)
4 cups (c.)	= 1 quart (qt.)
4 quarts (qt.)	= 1 gallon (gal.)
16 ounces (oz.)	= 1 pound (lb.)
1 kilogram (kg)	= 2.2 pounds (lb.)
28 grams (g)	= 1 ounce (oz.)
454 grams (g)	= 1 pound (lb.)
240 milliliters (mL)	= 1 cup (c.)
30 milliliters (mL)	= 1 fluid ounce (fl. oz.)
15 milliliters (mL)	= 1 tablespoon (T, Tbsp.)

2. Learn the equations in Table A-2 to convert temperatures on the Celsius scale (ºC) to temperatures on the Fahrenheit scale (°F) and vice versa.

Table A-2 Temperature Conversion using Celsius and Fahrenheit	
°C = (°F - 32)/1.8	°F = 1.8 (°C) + 32

3. Practice conversion of units by completing Table A-3 and filling in each blank with the equivalent weight, measurement, or temperature.

Table A-3 Unit Conversion Exercise	
_____ Tbsp. = 3 tsp.	1 c. = _____ Tbsp. = _____ fl. oz.
1 gal. = _____ fl. oz.	480 mL = _____ c.
1 oz. = _____ g	5 lb. = _____ kg
_____ qt. = 1.5 L.	10 Tbsp. + 2 tsp. = _____ c.
_____ c. = 2 gal.	1 kg = _____ oz.
_____ c. = 20 Tbsp.	_____ tsp. = ¾ c.
750 mL = _____ fl. oz.	240 mL = _____ pt.
66 °C = _____ °F	90 °F = _____ °C
22 °F = _____ °C	82 °C = _____ °F

B. *Measuring Techniques for Flour*

Objectives

1. To learn correct techniques for measuring flour.
2. To compare experimental results with standard results.

Basic Procedure to Measure Flour

Ingredients

4 c. flour

Procedure

1. Weigh an empty 1-c. measuring utensil (see Figure 2-1).
2. Spoon unsifted flour directly into the 1-c. measuring utensil.
3. Level flour with the straight edge of a spatula (see Figure 2-2).
4. Weigh the flour and the cup on the scale and record the combined weight.
5. Subtract the weight of the empty cup from the combined weight. Record the weight of the flour in Table B-1.
6. Add the flour weights obtained by the class and divide by the number of weights to obtain the class average; record the average weight in Table B-1.

Figure 2-1: Measuring container on a digital scale (obtaining tare weight).

Variations

1. Spooned, unsifted flour – Follow the basic procedure to measure flour as above.

2. Spooned, sifted flour – Follow the basic procedure to measure flour as above, except sift approximately 1 ½ c. of flour onto a flat surface. Then spoon the flour into a 1-c. measuring utensil.

3. Sifted flour – Follow the basic procedure to measure flour as above, except place a 1-c. measuring utensil on a flat surface and sift flour directly into the cup.

Table B-1 Measuring Techniques for Flour

Method	Weight of cup and flour	Weight of cup	Weight of flour	Class avg. weight of flour	Standard weight
Spooned, unsifted					4.5 oz. (125 g)
Spooned, sifted					4.3 oz. (~ 120 g)
Sifted					4.1 oz. (115 g)

Figure 2-2: Correct method for leveling off a 1-cup volume measure of flour.

C. Measuring Techniques for Sugar

Objectives
1. To learn correct techniques for measuring various sugar products.
2. To compare experimental results with standard results.

Basic Procedure to Measure Sugar

Ingredients
1 c. granulated sugar
2 c. brown sugar
2 c. confectioners' sugar

Procedure
1. Weigh an empty 1-c. measuring utensil.
2. Spoon granulated sugar into the 1-c. measuring utensil.
3. Level the sugar with the straight edge of a spatula.
4. Weigh the sugar and the cup on the scale and record the combined weight.
5. Subtract the weight of the empty cup from the combined weight. Record the weight of the sugar in Table C-1.
6. Add the sugar weights obtained by the class and divide by the number of weights to obtain the class average; record the average weight in Table C-1.

Variations

1. Granulated sugar – Follow the procedure to measure granulated sugar as above.

2. Brown sugar, packed – Follow the procedure to measure granulated sugar, except pack the brown sugar into a 1-c. measuring utensil with a spoon and level off with a straight edge.

3. Brown sugar, not packed – Follow the procedure to measure granulated sugar, except gently spoon the brown sugar into a 1-c. measuring utensil without packing it.

4. Confectioners' sugar, sifted – Follow the procedure to measure granulated sugar, except place a 1 c. measure on a flat surface and sift the confectioners' sugar directly into the measuring utensil.

5. Confectioners' sugar, unsifted – Follow the procedure to measure granulated sugar, except gently spoon the confectioners' sugar directly into the 1-c. measuring utensil.

Table C-1 Measuring Techniques for Sugar

Type of sugar	Weight of cup and sugar	Weight of cup	Weight of sugar	Class avg. weight of sugar	Standard weight
Granulated sugar					7.1 oz. (200 g)
Brown sugar, packed					7.1 oz. (200 g)
Brown sugar, unpacked					5.3 oz. (148 g)
Confectioners' sugar, sifted					3.4 oz. (95 g)
Confectioners' sugar, unsifted					4.2 oz. (118 g)

D. *Measuring Techniques for Fats*

Objectives
1. To learn correct techniques for measuring fats.
2. To compare experimental results with standard results.
3. To recognize weight differences for various forms of fat.

Basic Procedure To Measure Fat

Ingredients
Hydrogenated shortening

Procedure
1. Weigh an empty 1-c. measuring utensil.
2. Spoon hydrogenated shortening into the 1-c. measuring utensil.
3. Press the hydrogenated shortening into the cup to remove air. Level the hydrogenated shortening with the straight edge of a spatula.
4. Weigh the hydrogenated shortening and the cup on the scale and record the combined weight.
5. Subtract the weight of the empty cup from the combined weight. Record the weight of the hydrogenated shortening in Table D-1.
6. Add the shortening weights obtained by the class and divide by the number of weights to obtain the class average; record the average weight in Table D-1.

Variations

1. Hydrogenated shortening – Follow the procedure to measure fat as above.

2. Hydrogenated shortening, water displacement – Pour 1 c. cold water into a 2-c. graduated liquid measuring cup. Weigh. Spoon hydrogenated shortening into the measuring cup until the water level equals 2 c. Reweigh. Weigh plate. Remove hydrogenated shortening from measuring utensil, shake off excess water, and place the hydrogenated shortening on the plate. Reweigh the plate to determine the weight of the hydrogenated shortening.

3. Stick hydrogenated shortening – Weigh a plate. Unwrap and place a 1-c. stick of hydrogenated shortening on the plate. Reweigh the plate to determine the weight of the hydrogenated shortening.

4. Stick margarine – Weigh a plate. Unwrap and place a 1-c. stick of margarine on the plate. Reweigh the plate to determine the weight of the stick margarine.

5. Whipped margarine – Follow the procedure to measure fat as above, except substitute whipped margarine for hydrogenated shortening.

Table D-1 Measuring Techniques for Fats					
Type of fat	**Weight of utensil and fat**	**Weight of utensil**	**Weight of fat**	**Class avg. weight of fat**	**Standard weight**
Hydrogenated shortening					7 oz. (196 g)
Hydrogenated shortening, water displacement					7 oz. (196 g)
Stick hydrogenated shortening					7 oz. (196 g)
Stick margarine					8 oz. (224 g)
Whipped margarine					6 oz. (168 g)

E. Measuring Techniques for Liquids

Objectives

1. To learn correct techniques for measuring liquids.
2. To compare experimental results with standard results.
3. To become familiar with metric measurements for liquids.

Basic Procedure To Measure Liquids

Ingredients

Water

Procedure

1. Place a 1-c. graduated glass liquid measuring cup on a level surface.
2. Fill the cup with water to the 1 c. mark. Read the volume at the bottom of the meniscus.
3. Transfer the water from the measuring cup to a graduated cylinder. Read the volume at the bottom of the meniscus. Record the volume in milliliters in Table E-1. (See Figures 2-3 and 2-4.)
4. Add the water volumes obtained by the class and divide by the number of volumes to obtain the class average; record the average volume in Table E-1.

Variations

1. 1 cup – Follow the basic procedure to measure liquids as above.
2. ¾ cup – Follow the basic procedure to measure liquids, except measure ¾ c. water.
3. $^2/_3$ cup – Follow the basic procedure to measure liquids, except measure $^2/_3$ c. water.
4. ½ cup – Follow the basic procedure to measure liquids, except measure ½ c. water.
5. $^1/_3$ cup – Follow the basic procedure to measure liquids, except measure $^1/_3$ c. water.
6. ¼ cup – Follow the procedure to measure water as above, except measure ¼ c. water.

Figure 2-3: 10 mL measured in containers of different sizes.

Figure 2-4: Eye level for measuring liquids.

Table E-1 Measuring Techniques for Liquids			
Volume in glass cup	**Volume in graduated cylinder**	**Class average volume in graduated cylinder**	**Standard volume**
1 c.			237 mL
¾ c.			
⅔ c.			
½ c.			
⅓ c.			
¼ c.			

F. Using a Thermometer

Objectives

1. To correctly read and calibrate a thermometer.

Basic Procedure to Calibrate a Dial Stem Thermometer

Ingredients

Water
Ice

Procedure

1. Confirm that the dial stem thermometer can be calibrated. A dial stem thermometer with a hex nut under the thermometer dial is able to be calibrated.
2. Pack a beaker with crushed ice; add water to cover ice.
3. Place thermometer stem at least 3" in ice water and wait 2 minutes for the temperature to stabilize. The bulb should not touch the bottom or the sides of the beaker.
4. Thermometer should read 32 °F/0 °C. If the temperature is incorrect, use a pair of pliers to grip the hex nut and twist the dial face to display 32 °F/0°C.

G. Effect of Pan Surface Characteristics on Energy Transfer

Objectives

1. To demonstrate the influence of baking pan material on heating of a rich dough such as cookie dough.
2. To observe the effects of shiny and dull aluminum cookie sheets on the color and texture of cookies.
3. To observe the difference in baking in a convection and a conventional oven.

Basic Recipe for Sugar Cookies

Ingredients

1 ½ c. sugar
½ c. shortening
½ c. margarine
2 eggs
2 ¾ c. flour
2 t. cream of tartar
1 t. baking soda
¼ t. salt

Procedure

1. Preheat conventional oven to 400 °F.
2. Cream sugar, shortening, and margarine. Beat in the eggs.
3. Blend in a mixture of the flour, cream of tartar, soda, and salt.
4. Shape dough into small balls.
5. Place 4 to 5 of the small dough balls 2" apart on an ungreased, very dark cookie sheet.
6. Bake exactly 9 minutes at 400 °F. Transfer baked cookies to cooling rack.
7. Evaluate the appearance, flavor, and texture of the cookies; record observations in Table G-1.

Variations

1. Very dark cookie sheet, conventional oven – Follow the basic recipe for sugar cookies as above.

2. Light cookie sheet, conventional oven – Bake 4 to 5 of the cookies prepared by the basic recipe for sugar cookies on an ungreased, light-colored cookie baking sheet.

3. Air bake cookie sheet, conventional oven – Bake 4 to 5 of the cookies prepared by the basic recipe for sugar cookies on an ungreased air bake cookie sheet.

4. Glass pan, conventional oven – Bake 4 to 5 of the cookies prepared by the basic recipe for sugar cookies on an ungreased glass pan.

5. Silicone pan, conventional oven – Bake 4 to 5 of the cookies prepared by the basic recipe for sugar cookies on an ungreased silicone baking pan.

6. Very dark cookie sheet, convection oven – Bake 4 to 5 of the cookies prepared by the basic recipe for sugar cookies on an ungreased, very dark (Teflon) cookie sheet in a convection oven for exactly 9 minutes at 400 °F.

7. Light cookie sheet, convection oven – Bake 4 to 5 of the cookies prepared by the basic recipe for sugar cookies on an ungreased, light-colored cookie sheet in a convection oven for exactly 9 minutes at 400 °F.

8. Air bake cookie sheet, convection oven – Bake 4 to 5 of the cookies prepared by the basic recipe for sugar cookies on an ungreased air bake cookie sheet in a convection oven for exactly 9 minutes at 400 °F.

9. Glass pan, convection oven – Bake 4 to 5 of the cookies prepared by the basic recipe for sugar cookies on an ungreased glass pan in a convection oven for exactly 9 minutes at 400 °F.

10. Silicone pan, convection oven – Bake 4 to 5 of the cookies prepared by the basic recipe for sugar cookies on an ungreased silicone pan in a convection oven for exactly 9 minutes at 400 °F.

Table G-1 Effect of Pan Surface Characteristics on Energy Transfer

Conventional oven cookie characteristics		1. Dark pan (Teflon coated)	2. Light pan	3. Air pan	4. Glass pan	5. Silicone pan
Appearance	Top					
	Bottom					
Texture						
Flavor						
Convection oven cookie characteristics		**6. Dark pan (Teflon coated)**	**7. Light pan**	**8. Air pan**	**9. Glass pan**	**10. Silicone pan**
Appearance	Top					
	Bottom					
Texture						
Flavor						

H. Effect of Container Material on Energy Transfer

Objectives

1. To compare the transfer of energy of various materials.
2. To gain experience in plotting data.

Procedure to Monitor Decrease in Water Temperature

Ingredients

2 ¼ c. water

Procedure

1. Measure ¾ c. water into a small saucepan.
2. Bring the water to a full boil on top of the stove with the heat set to high.
3. As soon as the water comes to a boil, transfer the water to a Pyrex 1-c. liquid measuring cup.
4. Read the water temperature immediately after the transfer is made. Be sure your thermometer is suspended properly from a ring stand.
5. Continue to read and record the temperature at 30 sec intervals until room temperature is reached. Record the times and temperatures in Table H-1
6. Plot your data on Graph H-1. Let the Y axis represent temperature and let the X axis represent time.

Variations

1. Pyrex measuring cup – Follow the above procedure to monitor the decrease in water temperature.

2. Metal measuring cup – Follow the procedure to monitor the decrease in water temperature, except transfer ¾ c. boiling water to a metal measuring cup for temperature monitoring.

3. Styrofoam cup – Follow the procedure to monitor the decrease in water temperature, except transfer ¾ c. boiling water to a styrofoam cup for temperature monitoring.

Table H-1 Effect of Container Material on Energy Transfer

Pyrex		Metal		Styrofoam	
Time	Water temperature	Time	Water temperature	Time	Water temperature

Graph H-1

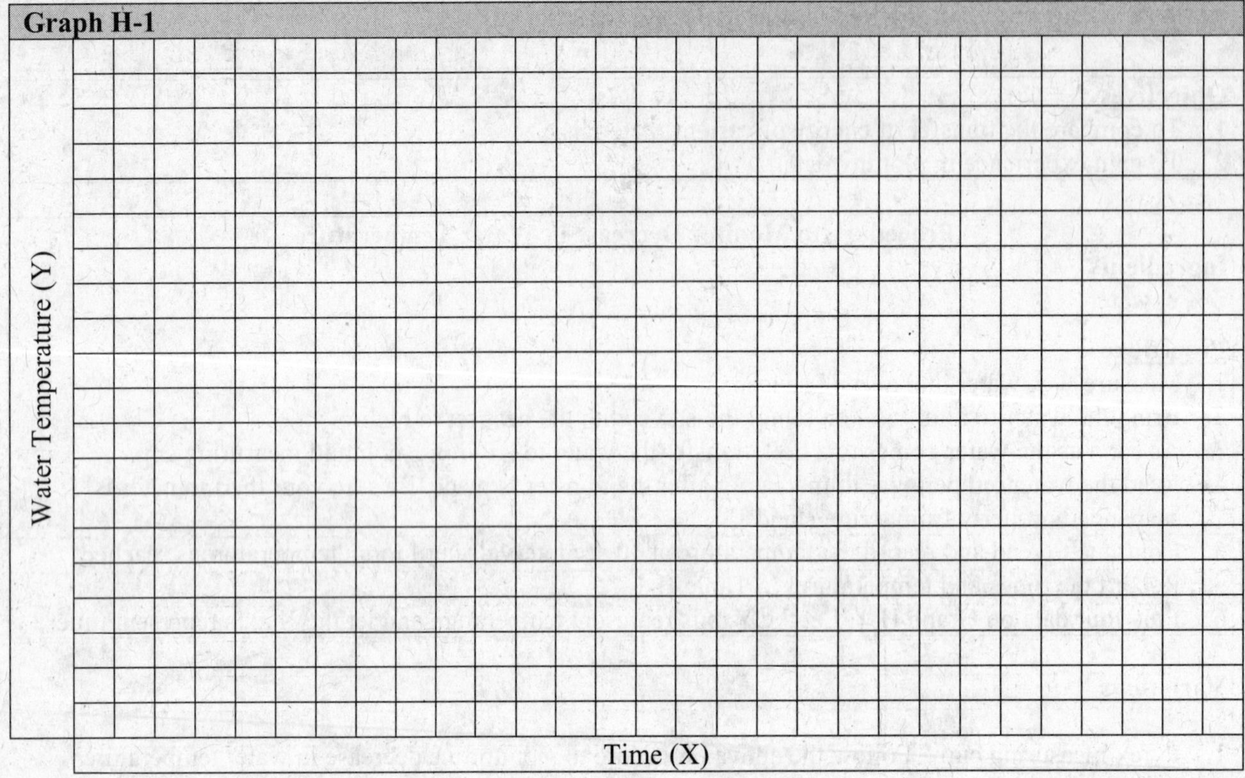

I. *Effect of Container Shape on Energy Transfer*

Objectives

1. To compare the transfer of energy in containers of the same material but with different shapes.
2. To gain experience in plotting data.

Basic Procedure to Observe Effect of Container Shape on Energy Transfer

Ingredients

4 c. water

Procedure

1. Preheat the oven to 400 °F. The oven rack should be in the center position.
2. Measure 2 c. room temperature water into each of the following: a small aluminum loaf pan and an 8" × 8" square aluminum cake pan.
3. Place both pans on the oven rack. Be sure the pans are positioned so that air flow around all sides of the pans is possible.
4. Read and record the initial water temperature. Close the oven door.
5. Read the water temperature at 4-minute intervals and record in Table I-1. Record the highest temperature observed. Do not keep the oven door open any longer than is necessary.
6. Plot your data on Graph I-1. Let the Y axis represent temperature and let the X axis represent time.

Table I-1 Effect of Container Shape on Energy Transfer

Aluminum Loaf Pan		Aluminum Square Pan	
Time	Temperature	Time	Temperature

Graph I-1

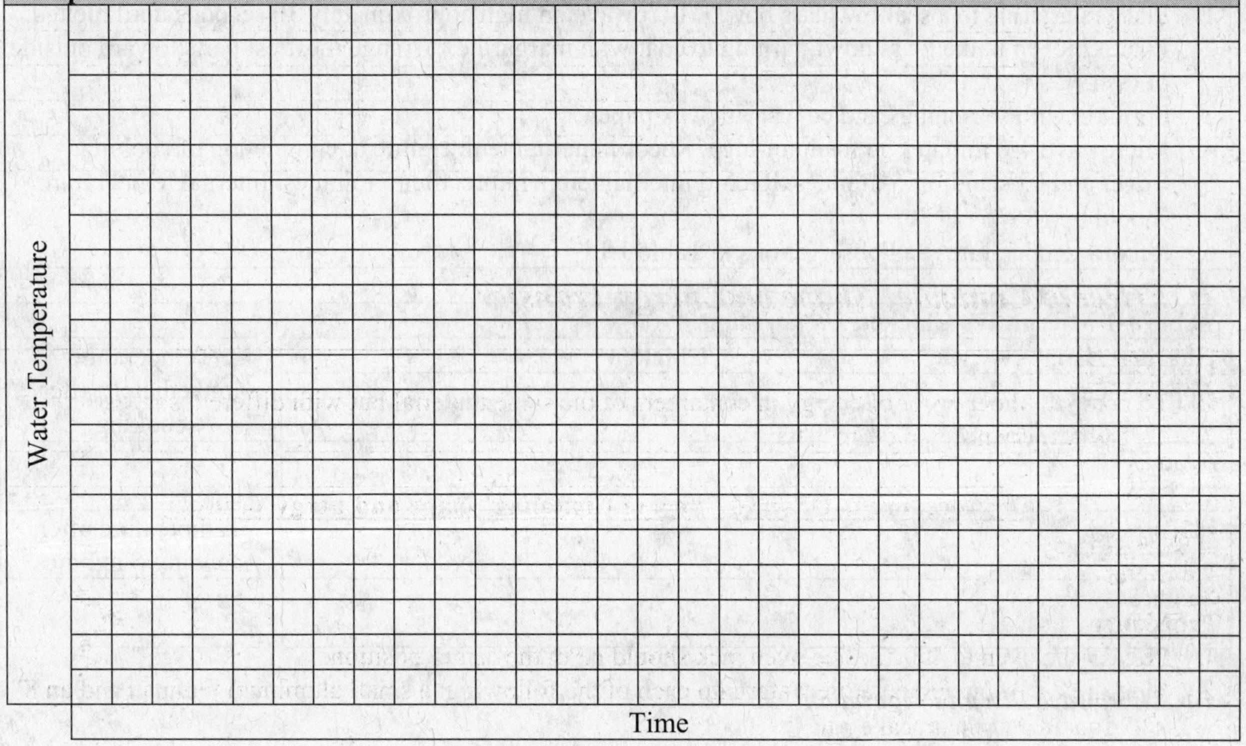

Water Temperature

Time

J. Microwave Energy Transfer

Objectives
1. To demonstrate how microwave energy heats food.
2. To utilize microwave cooking procedures for food preparation.
3. To observe palatability of products prepared in the microwave.

J-a. Microwave Chicken Preparation

Basic Recipe for Microwave Chicken

Ingredients
2 t. margarine
2 chicken filets (4 oz. each; refrigerated, not frozen)
~ ¼ t. paprika
~ ¼ t. salt
~ ¼ t. pepper

Procedure
1. Place margarine in a shallow glass bowl. Microwave on high approximately 30 seconds until melted.
2. Place chicken in the glass bowl, turning to coat with margarine. Arrange meatiest parts toward outside of dish.
3. Sprinkle with seasonings and cover with wax paper.
4. Microwave 4-6 minutes on medium-high. Record internal temperature at completion of cooking.
5. Cover and let stand for 5 minutes. Record internal temperature after 5 minutes. Internal temperature should be 165 °F.
6. Record temperatures and observations in Table J-1.

Table J-1 Microwave Chicken Preparation

Characteristic	Chicken	Temperature immediately after cooking
Appearance		
Aroma		
Flavor		Temperature after standing 5 minutes
Texture		

J-b. Microwave Biscuit Preparation

Basic Recipe for Microwave Biscuits

Ingredients
2 individual refrigerated canned biscuits

Procedure
1. Place 2 refrigerated canned biscuits on a microwave-safe plate.
2. Microwave 2 minutes on medium power, rotating dish a half-turn after 1 minute.
3. Cut 1 of the biscuits into fourths and taste immediately after cooking.
4. Allow remaining biscuit to stand for 2 minutes, cut into fourths and taste.
5. Record observations in Table J-2.

Table J-2 Microwave Biscuit Preparation		
Characteristic	**Biscuits immediately after cooking**	**Biscuits after standing**
Appearance		
Aroma		
Flavor		
Texture		

K. Reheating Using Microwave Energy Transfer

Objectives

1. To utilize microwave energy for reheating foods.
2. To observe palatability of products reheated in the microwave.

Basic Procedure for Reheating in the Microwave

Ingredients

2 baked dinner rolls

Procedure

1. Place 2 baked rolls 2" apart on a microwave-safe plate and place in the microwave oven.
2. Microwave 10 seconds. Cut one roll into fourths and taste immediately. Allow the other roll to sit for 2 minutes, cut into fourths and then taste.
3. Compare the overall palatability of the reheated rolls and record observations in Table K-1.

Variations

1. 10 seconds – Follow the basic procedure for reheating in the microwave as above.
2. 16 seconds – Follow the basic procedure for reheating in the microwave as above, except microwave rolls for 16 seconds.
3. 22 seconds – Follow the basic procedure for reheating in the microwave as above, except microwave rolls for 22 seconds.
4. 28 seconds – Follow the basic procedure for reheating in the microwave as above, except microwave rolls for 28 seconds.

Table K-1 Reheating Using Microwave Energy Transfer								
Characteristic	**10 sec. roll**		**16 sec. roll**		**22 sec. roll**		**28 sec. roll**	
	Immed. tasting	After 2 min.	Immed. tasting	After 2 min.	Immed. tasting	After 2 min.	Immed. tasting	After 2 min.
Appearance								
Aroma								
Flavor								
Texture								

L. *Defrosting Using Microwave Energy Transfer*

Objectives

1. To utilize microwave energy for defrosting foods.
2. To observe the quality of products defrosted in the microwave.

Basic Procedure for Defrosting in the Microwave

Ingredients

2 frozen ground beef patties (¼ lb. each)

Procedure

1. Place two ¼-lb. frozen ground beef patties on separate microwave-safe plates.
2. Microwave one patty on high for 4 minutes. Microwave the other patty for 4 minutes on defrost.
3. Observe the internal quality of the meat. Evaluate the extent of defrosting or cooking that has occurred and record observations in Table L-1. **Note: You will not taste this product.**

Table L-1 Defrosting Using Microwave Energy Transfer		
Characteristic	**Hamburger Patty, High Setting**	**Hamburger Patty, Defrost Setting**
Appearance		
Aroma		
Texture		

Post-Lab Study Questions

Textbook reference: Chapters 4 and 5

Discussion Questions

1. Which measuring utensils are best for measuring $^5/_8$ cup granulated sugar?

2. Which method of measuring flour volume gives results closest to the standard weight of all-purpose flour?

3. Give some reasons why the experimental weight of a cup of flour might differ from the standard weight.

4. The standard weight of sifted flour is 115 g. Weighing and measuring sifted flour three times (three replications) gives results of 125 g, 127 g, and 123 g. Would you consider these measurements accurate or precise or both?

5. Why is granulated sugar volume easier to measure precisely than flour volume?

6. Why is granulated sugar easier to measure more precisely than brown sugar?

7. If you were using light brown sugar in a recipe instead of granulated sugar, would you use more or less or the same volume?

8. What are the standard techniques for measuring plastic fats such as shortening?

9. Which measure of fat will have the greater volume: Melted or re-solidified after melting?

10. During which time interval did the greatest drop in temperature occur?

11. By what method of energy transfer does hot water heat Pyrex, metal, and Styrofoam cups?

12. How was energy being transferred to the water in the aluminum pans heated in the oven?

13. If the aluminum pans containing water had been covered, would you expect the water to heat more or less slowly? Why?

Questions for Post-Lab Writing

14. Your thermometer, when calibrated, reads 102 °C in boiling water and -1 °C in an ice bath. Explain how you would correct for your thermometer's inaccuracy in each of the following cases.

    a. You wish to concentrate a sucrose solution so that it boils at 110 °C.

    b. You wish to record the correct boiling point of a liquid that boils when your thermometer reads 96 °C.

    c. You wish to record the correct freezing point of a solution that freezes when your thermometer reads -5 °C.

15. Which cookies had the brownest bottom surface? The palest bottom surface?

16. Account for the color differences observed in the baked cookies.

17. Rank the cookie sheets according to their ability to transmit heat energy.

18. Which cookie sheet would you choose when baking sugar cookies? Why?

19. How did the Pyrex, metal, and Styrofoam cups influence the time-temperature data? How did the shape of the aluminum pan placed in the oven influence the time-temperature data? Which other variables might influence the data in this experiment?

20. Review your observations of the ground beef defrosted in the microwave. When defrosting foods in a microwave, why might a low setting be used instead of a high setting?

21. Based on your observations of the biscuits and chicken prepared in the microwave, can all foods be successfully prepared in a microwave oven? Specify what quality standards are not met using microwave cooking.

Unit 3 – Meat

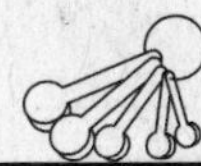

Introduction

Meat is defined as the flesh of animals used for food. It is derived primarily from muscle tissue. Tenderness is one of the most highly valued characteristics of meat. Differences in the amount and type of connective tissue present account for much of the variation in tenderness from muscle to muscle within the same animal. Muscles used for locomotion show an increase in connective tissue with a corresponding decrease in tenderness. Also, the amount of connective tissue in the muscle increases with the age of the animal. Enzymes that degrade both connective tissue and muscle proteins are sometimes applied to meat to increase tenderness. Proteases of plant origin, such as papain from papaya, ficin from figs, and bromelin from pineapple, may be used. These enzymes are activated as the temperature of the meat rises during cooking. Mechanical tenderization, carried out by passing pieces of meat through cutting blades, has a softening effect on less tender cuts. Varying cooking procedures also can influence tenderness. Moist-heat methods, including braising, stewing, and pressure cooking, are used for less tender cuts. Dry-heat methods, such as roasting, broiling, frying, and baking, are used for more tender cuts.

When heat is applied to meat, shrinkage occurs as a result of drip losses and evaporative losses. Drip losses refer to the materials left in the pan after cooking, such as fat, blood, and water. Evaporative losses refer to the water that evaporates from the meat drippings and from the meat itself while it is cooking. The higher the cooking temperature, the more muscle shrinks and the more evaporative losses there are. These losses may be calculated by weighing the meat before and after cooking. Varying the internal endpoint cooking temperature can alter the cooking losses and juiciness of the final product. A higher internal endpoint temperature for the meat usually leads to increased cooking losses and a less tender, less juicy product. Cooking meat to an internal temperature of 140 °F results in a rare piece of meat, while an internal temperature of 160 °F or 170 °F produces meat that is medium or well-done, respectively. Textured soy protein is often used as a meat extender. Addition of textured soy protein to a ground meat mixture reduces shrinkage during cooking because the fiber absorbs water and fats as they exit the meat.

Key Terms

au jus – natural juices or gravy from meat.

bake – to cook in an oven. Covered or uncovered containers may be used. When applied to meats in uncovered containers, it is called roasting.

bread – to coat foods by dipping into fine dry crumbs, into an egg-milk mixture, and then into crumbs again.

dredge – to sprinkle or coat with flour, sugar, or meal.

frizzle – to cook in a small amount of fat until crisp and curled at the edges, e.g., ham, bacon, dried beef.

fry – to cook in hot fat. When a small amount of fat is used, the process is known as pan-frying or sautéing or shallow-fat frying. When sufficient fat is used to keep food afloat, the process is known as deep-fat frying.

grill – to cook food by direct heat above an intense heat source.

lard – to cover uncooked lean meat or fish with strips of fat to enhance flavor and reduce dryness during cooking.

marinate – to cover with dressing or sauce and let stand for a length of time varying from less than an hour to overnight. Note: **Marinade** is the liquid or sauce in which the food is marinated.

score – to make shallow lengthwise or crosswise slits across the surface of food with a knife, fork, or other implement.

sear – to brown the surface of meat by a short application of intense heat in order to develop flavor and improve appearance.

stew – to simmer in a small amount of liquid.

Pre-Lab Questions

Textbook reference: Chapter 7

1. Identify the wholesale cuts of beef and name a retail cut that is derived from each of the wholesale cuts. Also indicate what cooking technique you would use for the retail cut.

2. What internal temperatures correspond to rare, medium, and well-done beef?

3. Identify the grades of beef. Compare choice and lean grades of beef.

4. What is the active ingredient in commercial tenderizers? How do commercial tenderizers function to tenderize meat?

5. What is marbling? How does it impact the quality of meat?

Lab Procedures

A. Comparison of Beef, Veal, Pork, and Lamb

Objectives

1. To determine similarities and differences among similar cuts of beef, pork, veal, and lamb.
2. To become familiar with the color and texture of the muscle, fat, and bone of beef, pork, veal, and lamb when the meat is raw.
3. To become familiar with the color and texture of the muscle, fat, and bone of beef, pork, veal, and lamb when the meat is cooked.

Basic Procedure to Compare Meat Products

Ingredients

Analogous cuts of beef, pork, veal, and lamb with bones in (chosen by instructor)

Procedure

1. Divide each type of meat into 2 portions.
2. Display one portion of each cut of meat for evaluation.
3. Broil (on the second rack from the top of the oven chamber) the other portion of each cut of meat to the medium-done stage (160 °F) and display cooked meat.
4. Examine the raw and cooked portions of meat. Record observations of each type of meat in Table A-1.

Figure 3-1: Portioning stew meat.

Characteristics of high-quality meat:

- Appearance: full brown color; slightly moist surface.
- Interior: medium doneness: light gray with slight pink in the center of the cut.
- Tenderness: moist mouth feel, slight resistance to chew; requires 5 to 6 chews to prepare for swallowing.
- Flavor: mild umami flavor.

Table A-1 Comparison of Beef, Veal, Pork, and Lamb

Meat	Muscle		Fat		Bone	
	Color	Texture	Color	Texture	Color	Texture
Beef, raw						
Beef, cooked						
Veal, raw						
Veal, cooked						
Pork, raw						
Pork, cooked						
Lamb, raw						
Lamb, cooked						

B. Comparison of Lean and Choice Beef

Objectives

1. To illustrate the differences in appearance between beef that is graded "choice" versus beef that is graded "lean" or "select."
2. To illustrate the difference in tenderness between choice versus lean or select beef.

Basic Recipe for Select and Choice Beef

Ingredients

4 oz. strip sirloin steak

Procedure

1. Examine steak. Note the color, amount, and location of the fat and connective tissue, and the color and grain of the muscle.
2. Record the raw weight of steak and additional data on the cooking record (Appendix A).
3. Record beginning cook time and cook the steak on a broiler pan that is 3"-5" below the broiler or on an electric grill that is set according to manufacturer directions.
4. Measure the temperature of the steak at the geometric center of the thickest portion.
5. Cook side one of the steak to internal temperature of 95 °F.
6. Flip steak and cook other side to internal temperature of 158 °F. Record end cooking time and final weight.
7. Place steak on plate, cover with foil, properly label, and place in holding oven.
8. Evaluate the tenderness (number of chews before swallowing) and juiciness of each piece of meat and record observations in Table B-1.

Variations

1. Select beef – Cook steak that is graded "Lean" or "Select" according to basic recipe and record observations in Table B-1.

2. Choice beef – Cook steak that is graded "Choice" according to basic recipe and record observations in Table B-1.

Table B-1 Comparison of Lean and Choice Beef

Meat	Description				Tender-ness (# of chews)	Juiciness	Cost per pound
	Muscle	Fat	Bone	Connective Tissue			
Choice							
Lean/Select							

C. Comparison of Connective Tissue in Meat and Cooking Methods

Objectives
1. To illustrate the amount and type of connective tissue present in meat.
2. To illustrate the response of connective tissue to dry- and moist-heat cooking methods.

Basic Recipe to Grill Strip Steak

Ingredients
4 oz. strip steak

Procedure
1. Examine cut of strip steak. Note the color, amount, and location of the fat and connective tissue, and the color and grain of the muscle.
2. Record the raw weight of steak and additional data on the cooking record (Appendix A).
3. Record beginning cook time and cook the steak on a broiler pan that is 3"-5" below the broiler or on an electric grill that is set according to manufacturer directions.
4. Measure the temperature of the steak at the geometric center of the thickest portion.
5. Cook side one of the steak to internal temperature of 95 °F.
6. Flip steak and cook other side to internal temperature of 158 °F. Record end cooking time and final weight.
7. Place steak on plate, cover with foil, properly label, and place in holding oven.
8. Evaluate the tenderness (number of chews before swallowing) and juiciness of each piece of meat and record observations in Table C-1.

Variations

1. Brisket strip – Prepare brisket strip steak according to the basic recipe to grill steak as above.

2. Loin strip – Prepare loin strip steak according to the basic recipe to grill steak as above.

Table C-1 Comparison of Connective Tissue in Meat and Cooking Methods						
Type of Meat	**Appearance**	**Juiciness**	**Tenderness (# of chews)**	**Flavor**	**Total Cook Time**	**% Yield***
Brisket strip						
Loin strip						

* % yield = Cooked weight / Original weight × 100

D. Effect of Tenderizers on Meat

Objectives

1. To demonstrate the use of enzymatic and mechanical methods of tenderizing connective tissue in meat.
2. To evaluate the relative tenderness and overall palatability of meats prepared with tenderizing agents.

Basic Recipe for Broiling Round Steak

Ingredients

4 oz. round steak

Procedure

1. Record the raw weight of steak and additional data on the cooking record (Appendix A).
2. Record beginning cook time and cook the steak on a broiler pan that is 3"-5" below the broiler or on an electric grill that is set according to manufacturer directions.
3. Measure the temperature of the steak at the geometric center of the thickest portion.
4. Cook side one of the steak to internal temperature of 95 °F.
5. Flip steak and cook other side to internal temperature of 158 °F. Record end cooking time and final weight.
6. Place steak on plate, cover with foil, properly label, and place in holding oven.
7. Evaluate the appearance, flavor, tenderness (number of chews before swallowing), and juiciness of each piece of meat.
8. Record observations in Table D-1.

Variations

1. No tenderizer treatment – Broil steak according to basic recipe above.
2. Mechanical tenderization – Mechanically tenderize one piece of the steak using a jacquard on both sides of the meat, paying special attention to areas containing apparent connective tissue. Broil steak according to basic recipe above.
3. Commercial meat tenderizer – Sprinkle ½ t. commercial meat tenderizer over the entire surface of a second piece of steak. Perforate the surface with a table fork, let stand 15 minutes, and then broil steak according to basic recipe above.

4. Marinade – Mix marinade of ½ c. vinegar and ¼ t. salt and marinate meat for 15 minutes. Broil steak according to basic recipe above.

5. Fresh pineapple – Peel and de-core fresh pineapple. Pulverize fresh pineapple in food processor. Place pulverized pineapple on steak and marinate for 15 minutes. Broil steak according to basic recipe above.

Table D-1 Comparison of Marinated Round Steak Products

Type of Treatment	Appearance	Juiciness	Tenderness (# of chews)	Flavor	Total Cook Time	% Yield*
No treatment						
Mechanical tenderizer						
Commercial meat tenderizer						
Vinegar and salt marinade						
Pulverized fresh pineapple						

* % yield = Cooked weight / Original weight × 100

E. Comparison of Beef Patty Products

Objectives

1. To illustrate the differences in preparation among various patty products.
2. To illustrate the differences in quality and sensory attributes among various patty products.

Basic Recipe for Beef Patties

Ingredients

¼-lb. ground beef patty

Procedure

1. Note and record raw weight, temperature, and beginning cooking time for each type of patty on the cooking record form (Appendix A).
2. Cook each patty in electric skillet at 350 °F to internal temperature of 95 °F.
3. Flip patty and cook other side to internal temperature of 165 °F.
4. Remove product from skillet and note cooked weight, temperature, and end cooking time on cooking record.
5. Record data and observations in Table E-1.

Variations

1. 73% ground beef – Prepare a ¼-lb. patty using 73% ground beef according to the basic recipe for beef patties.

2. Ground turkey – Prepare a ¼-lb. patty using ground turkey according to the basic recipe for beef patties.

3. 90% lean ground sirloin – Prepare a ¼-lb. patty using 90% lean ground sirloin according to the basic recipe for beef patties.

4. Boca® burger – Prepare a Boca burger patty according to the manufacturer's instructions.

5. Beef patty containing soy protein – Prepare a beef patty containing soy protein according to the basic recipe for beef patties.

6. Fully-cooked hamburger patty – Prepare a fully-cooked hamburger patty according to the manufacturer's instructions.

Table E-1 Hamburger Patty Comparison

Type of Meat	Appearance	Texture	Flavor	Total Cook Time	% Yield*	Cost per Pound
73% ground beef						
Ground turkey patty						
90% lean ground sirloin						
Boca® burger						
Beef patty w/ soy protein						
Fully-cooked hamburger patty						

* % yield = Cooked weight / Original weight × 100

F. Comparison of Protein Products Used in Stir Fry

Objectives:
1. To illustrate the usage of various proteins in food preparation.
2. To illustrate the differences in quality and sensory attributes among various sources of protein.

Basic Recipe for Stir Fry

Ingredients

1 T. oil
½ lb. protein source
1 c. frozen stir-fry vegetables
2 T. Worcestershire sauce
2 T. 2 t. soy sauce

Procedure

1. Heat 1 T. cooking oil in wok (see Figure 3-2) or electric skillet.
2. Thinly slice ½ lb. protein source and brown product, stirring quickly using only wooden utensils.
3. Add frozen stir-fry vegetables and stir-fry.
4. Add 2 T. Worcestershire sauce and 2 T. 2 t. soy sauce and cook until vegetables are tender.
5. Evaluate appearance, texture, and flavor. Record data and observations in Table F-1.

Figure 3-2: A wok is excellent for steaming or frying.

Variations

1. Beef – Prepare basic stir fry recipe as above using beef as protein source.
2. Pork – Prepare basic stir fry recipe as above using pork as protein source.
3. Firm tofu – Prepare basic stir fry recipe as above using firm tofu as protein source.

Table F-1 Comparison of Protein Products Used in Stir Fry				
Protein source	**Appearance**	**Texture**	**Flavor**	**Cost of protein source per pound**
Beef				
Pork				
Tofu				

Additional Food Preparation Exercises

G. Various Techniques for Meat Preparation

Objectives

1. To demonstrate a variety of ways to prepare meat.
2. To gain experience in preparing meat dishes.

G-a. Swiss Steak

Swiss Steak

Ingredients

1 lb. round steak
2 T. flour
2 T. vegetable oil
1 clove garlic
½ t. salt
¼ t. dry mustard
Pepper to taste
½ t. Worcestershire sauce
½ c. tomato juice
Chopped vegetables (e.g., onion, celery, carrot, green pepper, mushrooms) to taste
Additional flour (optional, for gravy)

Procedure

1. Cut the meat into approximately 4-oz. pieces and pound it with a meat hammer.
2. Rub the meat with the cut halves of the garlic clove.
3. Coat the meat with flour by shaking the meat and flour together in a plastic or paper bag.
4. Brown meat slowly on both sides in the vegetable oil over moderately low heat.
5. Add seasonings, vegetables and tomato juice.
6. Cover tightly and simmer slowly 1 ½ to 2 hours, or transfer to an ovenproof casserole and bake at 325°F. Turn meat once or twice during cooking.
7. At the end of cooking, the cooking liquid present may be thickened with a thin flour-cold water suspension that is mixed slowly into the cooking liquid while stirring with a wire whisk. Use approximately 2 T. flour for each cup of cooking liquid. With continuous stirring, bring the gravy to a full boil after adding the flour-water mixture. This ensures proper thickening and avoids any uncooked flour flavor.

G-b. Breaded Veal Cutlets

Breaded Veal Cutlets

Ingredients

1 slice veal cutlet
1 egg
1 T. water
½ c. bread crumbs
½ t. salt
Pepper to taste
¼ c. water
2 T. fat

Procedure

1. Remove any fat and bone from the veal.
2. Mix the slightly beaten egg with 1 T. water and coat the veal with egg, then with the bread crumbs seasoned with salt and pepper.
3. Sauté the breaded veal in the fat until golden brown.
4. Add the ¼ c. water; cover and simmer for 30 minutes.
5. Remove the cover and evaporate the water.

G-c. Oven-Braised Pork Chops

Oven Braised Pork Chops

Ingredients

½ lb. pork chops
2 T. flour
¼ t. salt
$^1/_8$ t. pepper
1 egg, beaten
1 T. milk
$^1/_3$ c. bread crumbs
2 T. oil

Procedure

1. Preheat oven to 350 °F.
2. Remove outer fat from pork chops.
3. Mix flour, salt, and pepper in a bag.
4. Add pork chops individually to the bag containing flour. Shake off excess flour while still in the bag.
5. Mix egg and milk together.
6. Spread bread crumbs on a sheet of waxed paper.
7. Dip floured pork chops into egg-milk mixture; then roll in bread crumbs.
8. Heat oil in a small skillet over high heat; do not allow oil to burn or smoke.
9. Add the chops and quickly brown on both sides.
10. Place browned chops in casserole dish. Add ¼ c. water.
11. Cover and place in oven. Bake 25-30 minutes.

G-d. Glazed Pork Tenderloin

Glazed Pork Tenderloin

Ingredients

2-3 lb. pork tenderloin
½ t. salt
1/8 t. pepper
4 T. honey
2 T. brown sugar
2 T. cider vinegar
1 t. spicy brown mustard

Procedure

1. Season pork tenderloin with salt and pepper. Place fat side up on cooking rack in a roasting pan.
2. Position a piece of aluminum foil over the top of the roaster pan to form a loose canopy. Do not completely cover the roasting pan. Bake at 325 °F for 1 hour.
3. Combine honey, brown sugar, cider vinegar, and mustard in a mixing bowl and mix well.
4. Spread glaze over tenderloin and bake uncovered for an additional hour or until it reaches an internal temperature of 155 °F.

From the recipe file of Karen Beathard

G-e. Braised (Pot) Roast

Braised (Pot) Roast

Ingredients

2-3 lb. beef chuck or arm roast, tripped of fat and connective tissue
1 T. vegetable oil
1 envelope dried onion soup mix
2 cans (10.5 oz.) cream of mushroom soup
Water
Aerosol vegetable oil cooking spray

Procedure

1. Place oil in skillet. Heat on medium for 4 minutes. Add roast to skillet and brown for 3-4 minutes on each side.
2. Spray 9" × 13" baking pan with aerosol vegetable oil cooking spray and transfer roast to prepared pan.
3. Rub onion soup mix on roast.
4. Mix 2 cans cream of mushroom soup and 2 cans of water in a mixing bowl. Pour soup mixture over roast.
5. Cover pan with foil and bake at 325 °F for 3-4 hours to an internal temperature of 145 °F.

From the recipe file of Alma Mynhier

Post-Lab Study Questions

Textbook reference: Chapter 7

Discussion Questions

1. Discuss the similarities and differences among cuts of beef, veal, pork, and lamb from the same part of the animal.

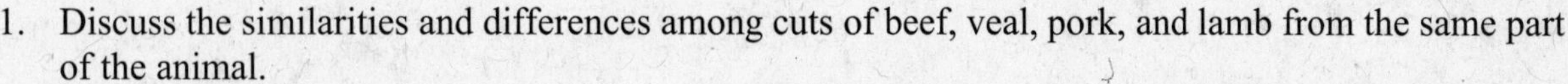

2. Explain how you would attempt to identify an unlabeled piece of meat.

Questions for Post-Lab Writing

3. Which tenderizing treatment was most desirable? Which was most effective?

4. When comparing tenderizing treatments, why is it so important to cook the meat to the same degree of doneness?

5. Fresh pineapple juice contains a proteolytic enzyme. Would canned pineapple juice be as effective as a tenderizing agent? Why?

6. What components of meat are included in evaporative loss? In drip loss?

Unit 4 – Poultry

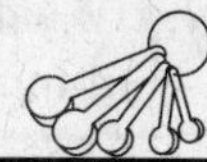

Introduction

Poultry are divided into market classes according to their species, age, and sex. The market classes for chicken include Cornish game hen, broiler or fryer, roaster, capon, and stewing chicken in order of age from youngest to oldest. Older poultry are less tender than young birds and require tenderization during cooking. The dark muscle of poultry usually is juicier, is more tender, and contains a higher fat content than the light poultry muscle. Poultry is a good source of high-quality protein. Its fat content varies with the market class. Most poultry can be satisfactorily cooked by dry-heat techniques although moist-heat techniques are also used to introduce variety.

Key Terms

baste – to moisten foods while cooking, especially while roasting meat. Melted fat, meat drippings, stock, water, or fruit juices may be used.

braise – to brown meat or vegetables in a small amount of fat, then to simmer slowly in a small amount of liquid in a covered container. The liquid may be juices from meat, or additional water, milk, meat stock, or vegetable juices.

fricassee – to cook by braising; usually applied to poultry or veal cut into pieces.

roasting – a dry-heat cooking method used primarily for meats and poultry. Roasted products are usually basted to maintain moisture and enhance flavor.

stir-fry – to fry by stirring and turning food very quickly and using a very small amount of fat.

truss – to bind or fasten together before cooking to enhance the maintenance of shape while cooking. Examples include skewering meat or binding wings of poultry before cooking.

Pre-Lab Questions

Textbook reference: Chapters 4 and 8

1. Which classifications of chickens are recommended for dry-heat cooking methods? Name five poultry dishes prepared using dry-heat methods.

2. What type of poultry is recommended for moist-heat cooking methods?

3. Compare the nutritional value of dark and light meat poultry products.

4. What is the minimum internal temperature to which poultry products should be cooked? Why is the minimum internal temperature for poultry set higher than that for beef or fish?

Lab Procedures

A. Comparison of Store Brand and National Brand Chicken

Objectives
1. To compare two brands of chicken for appearance, texture, and flavor.
2. To demonstrate a technique for preparation of chicken cutlets.

Basic Recipe for Chicken Cutlet Preparation

Ingredients
2 store brand boneless chicken breast halves
1 egg, beaten
½ c. dry bread crumbs
2 T. vegetable oil

Procedure
1. Flatten chicken breast halves, or pound them with a meat cleaver, until they are ~½ inch thick. Remove the skin from the chicken breasts.
2. Dip each piece of chicken in beaten egg and then into dry bread crumbs.
3. Place 2 T. vegetable oil in a skillet on medium-low heat.
4. Quickly dip 1 side of the coated chicken pieces in the oil. Immediately turn them over and cook the un-dipped sides in the remainder of the oil on medium heat for about 10 minutes.
5. When the first side is brown, turn the chicken over and continue to cook until the second side is brown and chicken is cooked to an internal temperature of 165°F. Do not cover skillet.
6. Evaluate appearance, texture, tenderness, and flavor of chicken and record observations in Table A-1.

Characteristics of high-quality breaded chicken:
- Appearance: golden brown, crunchy looking surface
- Tenderness: some resistance to chew
- Texture: crunchy, crisp surface; thick, firm center
- Flavor: mild meaty flavor, not greasy

Variations

1. Store brand chicken – Prepare according to basic recipe for chicken cutlet preparation as above.

2. National brand chicken – Prepare according to basic recipe for chicken cutlet preparation as above, except use name brand chicken breasts.

Table A-1 Comparison of Generic and National Brand Chicken				
Brand	**Appearance**	**Tenderness**	**Texture**	**Flavor**
Store brand chicken breast				
National brand chicken breast				

B. Turkey Deli Meat Comparison

Objectives

1. To compare turkey deli products for appearance, texture, flavor, and cost per pound.
2. To rank a series of samples based on overall quality.

Basic Procedure to Compare Turkey Deli Products

Ingredients

Least expensive sliced deli turkey
Mid-priced sliced deli turkey
Premium-priced sliced deli turkey

Procedure

1. Place 1-oz. deli turkey samples into 2-oz. sample cups marked with the respective sample codes. Prepare enough cups for each participant to sample all products.
2. Assess the quality characteristics of each sample including appearance, texture, aroma, and flavor.
3. Rank turkey samples in descending order of quality with the highest quality sample ranked #1 in Table B-1. You may re-taste any of the samples while ranking for quality; no ties are allowed in the ranking.
4. Record your observations of turkey quality and the cost per pound information for each sample in Table B-2.

Table B-1 Ranking of Turkey Deli Products	
Ranking Scale	**Rank Products Below**
#1 – Most preferred	
#2	
#3 – Least preferred	

Table B-2 Turkey Deli Meat Comparison				
Type of meat	**Appearance**	**Texture**	**Flavor**	**Cost per pound**

Additional Food Preparation Exercises

C. Various Techniques for Poultry Preparation

Objectives
1. To prepare poultry by various cooking methods.

C-a. Chinese Almond Chicken

Chinese Almond Chicken*

Ingredients

½ c. slivered almonds
2 uncooked chicken breast halves, skinless and boneless
1 T. oil
1 c. sliced onion
1 ½ c. diagonally-cut celery pieces
1 c. chicken broth
1 T. cornstarch
1 t. granulated sugar
¼ c. soy sauce
¾ c. chicken broth
5 oz. canned bamboo shoots, drained
5 oz. canned water chestnuts, drained and sliced
½ lb. fresh broccoli, cut into florets
¼ lb. fresh or frozen pea pods

Procedure

1. Cut across the grain of the chicken into very thin strips.
2. Heat the oil in a skillet over medium high heat.
3. Add the chicken to the skillet. Cook, stirring frequently, 5 to 10 minutes, or until just cooked through. Remove chicken.
4. Add sliced onion, celery slices, broccoli, pea pods, and ½ c. chicken broth. Cook uncovered 5 minutes, or until vegetables are slightly tender.
5. Combine the cornstarch, sugar, soy sauce, and ½ c. chicken broth. Pour over vegetables. Cook and stir until sauce thickens.
6. Add chicken, bamboo shoots, water chestnuts, and nuts. Mix well and heat thoroughly.

*One pound of turkey breast meat may be substituted for the chicken.

C-b. Chicken Divine

Chicken Divine*

Ingredients

20 oz. frozen broccoli spears
1 lb. cooked chicken breast, deboned, in bite-size pieces
2 cans (10.5 oz.) cream of mushroom soup
1 c. mayonnaise
1 t. lemon juice
½ t. curry powder
2 T. cooking sherry
1 c. medium cheddar cheese, shredded
30 round butter-flavored crackers, crushed
Aerosol vegetable oil cooking spray

Procedure

1. Prepare broccoli spears according to package directions.
2. Mix cream of mushroom soup, mayonnaise, lemon juice, curry powder, and cooking sherry in a mixing bowl.
3. Spray 11 ½" × 7 ½" × 2" baking pan with aerosol vegetable oil cooking spray. Arrange drained broccoli spears in prepared pan.
4. Place cooked chicken breast on top of broccoli and pour soup mixture over chicken.
5. Place crackers in a plastic bag and crush them. Add shredded cheddar cheese to crushed crackers, mix thoroughly, and spread over soup mixture.
6. Bake at 350 °F for 30 minutes to an internal temperature of 165 °F.

*May reduce fat content by substituting equivalent amounts of low-fat cream of mushroom soup, low-fat mayonnaise, and saltine crackers for cream of mushroom soup, mayonnaise, and butter crackers.

From the recipe file of Karen Beathard

C-c. Glazed Cornish Hen

Glazed Cornish Hen

Ingredients

1 Cornish hen (1-1 ½ pounds)
1 T. butter
2 T. apricot jam or orange marmalade
1 t. orange or cranberry juice
Dash dry mustard
Salt
Pepper

Procedure

1. Wash and dry the Cornish hen, inside and out.
2. Sprinkle the interior with the salt and pepper.
3. Rub the exterior with the butter.
4. Place the bird on the rack of a roasting pan. Tuck the wing tips behind the back of the bird.
5. Position a piece of aluminum foil over the top of the roaster pan to form a loose canopy. Do not completely cover the roasting pan.
6. Roast the bird at 350 °F for approximately 1 hour to an internal temperature of 165 °F.
7. Meanwhile, mix apricot jam or orange marmalade with orange or cranberry juice and dry mustard in preparation for basting.
8. About 20 minutes before the end of roasting, remove the foil canopy and baste the bird frequently with the jam-juice-dry mustard mixture.

C-d. Turkey and Broccoli Burgers

Turkey and Broccoli Burgers

Ingredients

10-oz. package frozen chopped broccoli
1 ½ lbs. lean ground turkey
1 c. extra sharp cheddar cheese, shredded
1 small onion, chopped
1 egg, beaten
¼ c. dried bread crumbs
1 T. Worcestershire sauce
1 t. salt

Procedure

1. Prepare broccoli according to package directions. Drain cooked broccoli well.
2. In a large mixing bowl, mix drained broccoli, ground turkey, shredded cheddar cheese, onion, egg, Worcestershire sauce, and salt.
3. Shape mixture into 8 patties (¾" thick).
4. Broil patties on greased baking pan for approximately 10 minutes. Turn patties over and broil on opposite side approximately 8 minutes to an internal temperature of 165 °F.

From the recipe file of Milayna Brandon

C-e. Chicken Spaghetti

Chicken Spaghetti

Ingredients

Whole chicken
1 T. salt
Water
2 celery stalks, diced
1 medium onion, diced
1 green bell pepper, diced
Garlic clove, minced
1 T. chili powder
½ t. oregano
1 can (10.5 oz.) cream of mushroom soup
2 cans tomato sauce (8 oz. each)
1 c. medium cheddar cheese, grated
13.25 oz. spaghetti
Aerosol vegetable oil cooking spray

Procedure

1. Remove entrails from chicken. Place chicken in a large pot and add enough water to cover it. Bring chicken to a boil, reduce temperature to simmer, and cook to an internal temperature of 165 °F.
2. Remove chicken and place it in a pan to cool.
3. Add ½ c. water, celery, onion, and green bell pepper to chicken broth and simmer for 20 minutes.
4. Prepare spaghetti according to package directions.
5. Add garlic, chili powder, oregano, mushroom soup, and tomato sauce to broth mixture and mix well.
6. Add cooked spaghetti and deboned, chopped chicken to mixture.
7. Spray a 9" × 13" baking dish with aerosol vegetable oil cooking spray. Pour spaghetti mixture into prepared pan. Cover mixture with cheddar cheese.
8. Place in oven at 350 °F to melt cheese.

From the recipe file of Alma Mynhier

C-f. Chicken Salad

Chicken Salad

Ingredients

4 chicken breasts or 1 whole chicken
Water
¼ c. celery, chopped
1 c. seedless red grapes, halved
1 can mandarin oranges
1 small can crushed pineapple
1 red apple, cut up
½ c. slivered almonds
½ c. mayonnaise
¼ c. poppy seed dressing

Procedure

1. Remove entrails from chicken. Place chicken in a large pot and add enough water to cover it. Bring chicken to a boil, reduce temperature to simmer, and cook to an internal temperature of 165 °F.
2. Remove chicken and place it in a pan to cool.
3. Mix celery, grapes, mandarin oranges, pineapple, apple, and almonds together in a mixing bowl.
4. Debone chicken (if using whole chicken). Cut cooked chicken into bite-size pieces. Mix chicken with fruit mixture.
5. Mix mayonnaise and poppy seed dressing in a separate bowl. Pour dressing over chicken mixture and mix well. Refrigerate until service.

From the recipe file of Nancy Morse

Post-Lab Study Questions

Textbook reference: Chapter 8

1. Compare the chicken cutlets prepared from the name brand and the store brand chicken. Was one superior? If so, cite possible reasons.

2. Describe the quality differences among the deli turkey samples. How did the price of the turkey compare to its quality?

Unit 5 – Fish and Shellfish

Introduction

Seafood is usually classified as finfish (vertebrate) and shellfish (invertebrate). Finfish may be further categorized as lean or fat. Lean fish, such as cod, flounder, haddock, snapper, and sea trout, contain less than 5% fat. Mullet, mackerel, lake trout, bluefish, salmon, tuna, and other fat fish have a fat content between 5 and 20%. Shellfish may be classified as mollusks, which have a very hard calcified shell covering a soft body, and crustaceans, which have an armor-like segmented outer shell. Clams, oysters and mussels are examples of mollusks, while shrimp, crabs, and lobsters are examples of crustaceans. Fish may be marketed as whole, drawn, dressed, steaks, fillets, and other forms. A drawn fish has only the entrails removed, while a dressed fish also has the head, tail, and fins removed. Fish steaks are cross-section slices of a dressed fish. Fillets are sections of the fish flesh cut parallel to the backbone. The muscle fibers of fish are short and their ends are inserted into sheets of connective tissue. During cooking, the collagen in the connective tissue is hydrolyzed to gelatin and the muscle segments separate as flakes, resulting in the characteristic "flakiness" of cooked fish. Since smaller amounts of connective tissue are present in fish muscle than in meat muscle, cooking methods that tenderize connective tissue are not required for fish. Fish proteins can be coagulated by heat, using such methods as broiling, baking, frying, steaming, and poaching.[1] Surimi is a product made from fish protein that has been colored and flavored to resemble products such as crab, shrimp, and lobster.

Key Terms

broil – to cook food by direct heat under an intense heat source.

fillet – a boneless piece of meat, fish, or poultry.

flake – to separate in small pieces. This is often used as an indicator of the doneness of fish.

kippered – salted and smoked (fish). Salmon and herring are often prepared using this technique.

poach – to cook in simmering hot liquid.

sauté – to fry using a very small amount of fat.

simmer – to cook in liquid at a temperature just below the boiling point (185 °F to 210 °F).

Pre-Lab Questions

Textbook reference: Chapter 9

1. Why is a short cooking time recommended for finfish and shellfish?

2. Why do fish recipes often include citrus juice, wine, or tomatoes?

[1] For tips on microwave preparation of fish, see http://www.simplyseafood.com/newsletters/10_06/zap.html.

3. Compare the amount of connective tissue in fish with the amount of connective tissue in poultry.

4. What are the ingredients in surimi?

5. What is ceviche?

Lab Procedures

A. Coagulation of Fish Protein by Heat—Comparison of Fresh and Frozen Fish

Objectives

1. To demonstrate the coagulation of fish proteins by broiling and poaching.
2. To compare the appearance, texture, and flavor of fresh and frozen fish fillets cooked by broiling and poaching.

Basic Recipe for Fish Preparation

Ingredients

4 oz. flounder fillet
1 t. melted butter
Aerosol vegetable oil cooking spray

Procedure

1. Note and record raw weight, temperature, and beginning cooking time for fish on the cooking record form (Appendix A).
2. Place 4 oz. fresh flounder fillets (or other white fish) on a foil-lined broiler pan that has been sprayed with aerosol vegetable oil cooking spray.
3. Place the pan 3" below the broiler. Brush the surface of the fish with melted butter.
4. Broil just 4-5 minutes or until cooked through. It is not necessary to turn the fish. The fish is done when the flesh becomes opaque and separates into flakes. Do not overcook.
5. Reweigh final fish product. Evaluate the appearance, texture, and flavor of the fish and record your assessment in Table A-1.

Characteristics of high-quality fish:

- Appearance: interior meat is white, shiny, moist, and opaque
- Texture: flakes pull off when a fork is inserted
- Tenderness: moist, slight resistance to chew

Variations

1. Broiled fresh fish – Broil 4 oz. fresh fish according to basic recipe for fish preparation.

2. Broiled thawed, previously frozen fish – Broil 4 oz. thawed, previously frozen fish according to basic recipe for fish preparation.

3. Microwaved fresh fish – Follow basic recipe for fish preparation as above, except microwave the fresh fish skin side down on HIGH in a covered microwave baking dish for approximately 3-6 minutes per pound in a 600- to 750-watt oven. Allow fish to remain on countertop for 5 minutes to complete cooking. Evaluate the appearance, texture, and flavor of the microwaved fish in Table A-1.

4. Microwaved thawed, previously frozen fish – Follow basic recipe for fish preparation as above, except microwave the thawed, previously frozen fish skin side down on HIGH in a covered microwave baking dish for approximately 3-6 minutes per pound in a 600- to 750-watt oven. Allow fish to remain on countertop for 5 minutes to complete cooking. Evaluate the appearance, texture, and flavor of the microwaved fish in Table A-1.

5. Poached fresh fish – Follow basic recipe for fish preparation as above, except poach the fresh fish by wrapping and tying it in a double layer of cheesecloth. Place the wrapped fish carefully into boiling salted water. (Use 2 t. salt per qt. of water.) At once, reduce the heat to simmering and simmer about 10 minutes or until fish flakes. Remove the fish from the liquid as soon as it is cooked. Evaluate the appearance, texture, and flavor of the poached fish and record your assessment in Table A-1.

6. Poached thawed, previously frozen fish – Follow basic recipe for fish preparation as above, except poach the thawed, previously frozen fish by wrapping and tying it in a double layer of cheesecloth. Place the wrapped fish carefully into boiling salted water. (Use 2 t. salt per qt. of water.) At once, reduce the heat to simmering and simmer about 10 minutes or until fish flakes. Remove the fish from the liquid as soon as it is cooked. Evaluate the appearance, texture, and flavor of the poached fish and record your assessment in Table A-1.

Table A-1 Fish Comparison

Type of Fish	Appearance	Texture	Flavor	Total Cooking Time	% Yield*	Cost per Pound
Broiled fresh fish						
Broiled thawed, previously frozen fish						
Microwaved fresh fish						
Microwaved thawed, previously frozen fish						
Poached fresh fish						
Poached thawed, previously frozen fish						

* % yield = Cooked weight / Original weight × 100

Additional Food Preparation Exercises

B. Selected Seafood Dishes

Objectives
1. To demonstrate various techniques for cooking seafood.
2. To gain practice in seafood preparation.
3. To become familiar with surimi.

B-a. Shrimp Bourgeoise

Shrimp Bourgeoise

Ingredients
1 lb. fresh or frozen shrimp
2 T. butter
½ clove garlic, minced
1 T. chopped fresh parsley
$^1/_8$ t. salt
Dash of cayenne pepper
Dash of black pepper
1 T. dry white wine
½ c. canned stewed tomatoes

Procedure
1. Melt butter in a saucepan over medium heat. Add garlic, cayenne, parsley, salt, and pepper and sauté gently for 3 minutes.
2. Add the shrimp and sauté lightly for 5 minutes.
3. Add wine and tomatoes and simmer gently 8 minutes, or until shrimp are tender. Do not overcook.

B-b. Surimi Creole

Surimi Creole

Ingredients
1 T. onion, minced
1 T. green pepper, minced
8-10 mushrooms, sliced
2 T. fat or oil
2 c. canned tomatoes, chopped
½ t. chili powder or pepper
½ t. salt
¾ lb. surimi, cut into bite-size pieces
½ c. uncooked rice
Water

Procedure
1. Cook the rice according to package directions.
2. Cook onion, green pepper, and mushrooms in fat over low heat.
3. Add tomatoes, chili powder or pepper, and salt.
4. Simmer until sauce thickens.
5. Add surimi and simmer 8-10 minutes.
6. Serve over cooked rice.

B-c. Surimi Cocktail

Surimi Cocktail

Ingredients
1 lb. surimi, cut into bite-size pieces
2 c. seafood cocktail sauce

Procedure
1. Arrange the surimi on a plate.
2. Using toothpicks, dip the surimi into the cocktail sauce, and eat. (The surimi is pre-cooked, so it can be eaten cold.)

B-d. Crayfish Cornbread

Crayfish Cornbread

Ingredients
2 eggs
1 t. salt
1 medium onion, chopped
½ c. vegetable oil
1 c. yellow corn meal
1 c. medium cheddar cheese, shredded
½ fresh jalapeño, chopped
16 oz. cream corn
1 lb. crayfish tails
Aerosol vegetable oil cooking spray

Procedure
1. Preheat oven to 375 °F.
2. Mix all ingredients together in a large mixing bowl.
3. Pour into a greased (with cooking spray) 2-qt. rectangular pan. Bake for approximately 30 minutes to an internal temperature of 165 °F.
4. Cool before cutting into portions.

B-e. Oven-Roasted Salmon with Caper Sauce

Oven-Roasted Salmon with Caper Sauce

Ingredients

1 ½ lb. salmon fillet
Salt and freshly ground black pepper
Extra virgin olive oil
¼ c. Hellmann's® Real Mayonnaise or sour cream
¼ c. nonfat plain yogurt
1 T. freshly squeezed lemon juice
¼ t. Tabasco® sauce or to taste
¼ - ½ t. Worcestershire sauce
3 T. brine-cured capers, rinsed and chopped
1 t. Dijon mustard

Procedure

1. Preheat oven to 425 °F.
2. Line a sheet pan with foil; place salmon on sheet pan. Season salmon fillet with salt and pepper and drizzle with olive oil.
3. Roast in preheated oven for 12-15 minutes (cooking time will vary depending on thickness of fillet).
4. In the meantime, combine next seven ingredients in a small bowl. Caper sauce may be prepared several hours in advance; refrigerate until serving.
5. Serve oven-roasted salmon with caper sauce on the side.

Post-Lab Study Questions

Textbook reference: Chapter 9

1. What market forms of fresh and frozen finfish are available?

2. Compare the fish cooked by broiling, poaching, and microwaving. What advantages and disadvantages can be seen for each type of cooking?

3. Compare the palatability of fresh fish versus frozen fish. What factors influence the quality of frozen fish?

4. How did the surimi compare with real crab and shrimp?

Unit 6 – Milk

Introduction

Milk is used widely in food preparation. It is a means of incorporating moisture into batters and dough and is an important ingredient in prepared foods such as cream soups, sauces, puddings, and milk foams. It is also the raw material for fermented milk products such as yogurt and sour cream.

There are many types of milk available. Commercial buttermilk is produced by adding a bacterial culture to fluid milk, which converts some of the milk's lactose to lactic acid. "Real buttermilk" is the liquid remaining when cream is churned and the resulting butter removed. Acidophilus milk is produced by adding a culture of *Lactobacillus acidophilus* to pasteurized milk. Evaporated milk is fresh fluid milk from which fifty percent of the water has been removed. It is then thermally processed to produce a sterile product. Sweetened condensed milk is a concentrated milk product that contains 44% sucrose. It is processed with a less severe heat treatment than that used for evaporated milk.

The two major classes of milk proteins are caseins and whey or serum proteins. Casein is affected little by heat, but is affected greatly by the addition of acid. The casein molecule has a net negative charge, which is neutralized by the addition of acid. This causes the casein micelles to become destabilized and coagulate. Whey or serum proteins are affected little by acid, but are affected greatly by the addition of heat. The application of heat removes the water shell, which normally keeps the whey proteins stabilized. These denatured proteins precipitate to the bottom of the pan when milk is heated and may cause scorching.

Milk heated in an open container will develop a skin on the surface. Evaporation of water from the heated milk results in a concentration of casein, milk fat and calcium, and phosphate salts at the surface, causing the skin to form.

Foods that contain acids may cause milk products to thicken or curdle, as in tomato soup, scalloped potatoes, and creamed dishes. Flour added to these dishes acts to protect the milk proteins from the acids.

Milk may be coagulated by the action of the enzyme chymosin on the casein micelle. Chymosin, commonly derived from genetically modified bacteria, is often used to coagulate milk for puddings and cheeses. Chymosin extracted from the stomachs of ruminant animals (rennet) may also be used.

Key Terms

casein – major protein in milk that denatures and coagulates with exposure to an acidic environment or enzyme exposure.

chill – to place in a refrigerator or cool place until cold.

coagulation – the conversion of a liquid to a soft semisolid mass.

curd – the coagulated component of milk that results from the denaturation and coagulation of casein.

reconstitute – to restore a dehydrated product to liquid form through the addition of water. Nonfat dry milk is reconstituted for consumption.

roux – a mixture of butter and fat that is cooked and used to thicken sauces, gumbo, soups, and gravy.

scald – to heat liquid, usually milk, until hot but not boiling.

scallop – to bake food, usually cut in pieces, with a milk-based sauce or other liquid.

whey – protein in milk that denatures in response to the application of heat. It is also the watery portion of milk removed from the curd in cheese production that consists of water, lactose, and whey protein.

white sauce – sauce comprised of fat, flour, milk or stock, and seasoning that is often served with vegetables, fish, or meat.

Pre-Lab Questions

Textbook reference: Chapter 10

1. Which protein in milk is most affected by heat? Which protein is most affected by acid?

2. Explain the differences among whole milk, evaporated milk, condensed milk, and buttermilk.

3. What steps would you take to prepare a stable milk-based foam for a pie filling?

4. Will heavy whipping cream, evaporated milk, or reconstituted non-fat milk solids result in a foam with the largest volume? Why? How will the stability of these foams compare?

5. What process is used to make yogurt?

Lab Procedures

A. Sampling of Milk Products

Objectives

1. To compare appearance, consistency, flavor, aroma, and composition of various commercial milk products.
2. To become familiar with some of the milk products available on the market.

Basic Procedure to Sample Milk Products

Ingredients

Assortment of milk products (e.g., whole, low-fat, nonfat, lactase-treated, etc.; chosen by instructor)
2-oz. sample cups

Procedure

1. Pour approximately 1 T. of each milk product into 2-oz. sample cups. Pour enough cups for each participant to sample each product.
2. Observe the cost, appearance, aroma, flavor, and consistency of each milk product and record observations in Table A-1.
3. Read the labels on the milk containers and record differences in the composition of the various products in Table A-1.

Table A-1 Sampling of Milk Products

Milk product	Cost	Appearance	Aroma	Flavor	Consistency	Composition

B. *Sampling of Yogurt Products*

Objectives

1. To compare appearance, consistency, flavor, aroma, and composition of various yogurt products.
2. To become familiar with some of the yogurt products available on the market.

Basic Procedure to Sample Yogurt Products

Ingredients

Assortment of yogurt products with different fat levels, live bacteria (probiotics), and flavors
2-oz. sample cups

Procedure

1. Follow the same sampling procedure used for milk products in Lab 6A.
2. Record observations and composition information for each yogurt product in Table B-1.

Table B-1 Sampling of Yogurt Products

Product	Cost	Appearance	Aroma	Flavor	Consistency	Composition

C. Coagulation of Fresh Whole Milk

Objectives

1. To observe and explain the effect of heat on fresh whole milk.
2. To observe and explain the effect of acid on fresh whole milk.

C-a. Effect of Heat

Basic Procedure to Evaluate the Effect of Heat

Ingredients

½ c. whole milk

Procedure

1. Put ½ c. whole milk into a saucepan and place over low heat. Do not cover the saucepan or stir or boil the milk.
2. Continue heating until a thick skin develops on the surface of the milk and a definite precipitate is visible on the bottom of the saucepan. Do not worry about overheating the milk. It needs no attention while heating.
3. Identify the milk component involved in the following changes, and record in Table C-1: formation of a film on the milk surface; precipitation on the bottom of the saucepan; and browning on the bottom of the saucepan.

C-b. Effect of Acid

Basic Procedure to Evaluate the Effect of Acid

Ingredients
1 c. whole milk
2 T. vinegar

Procedure
1. Measure 1 c. whole milk into a glass graduated liquid measuring cup.
2. Use pH paper to determine the pH of the milk and the vinegar to be used and record the values in Table C-2.
3. Add 1 t. vinegar to the milk. Stir well.
4. Let the milk stand for 2 minutes. Determine the pH of the milk and observe it for any thickening or curd formation. Record the pH and observations in Table C-2.
5. Repeat steps 3 and 4 five more times until a total of 2 T. vinegar has been added to the milk.
6. Identify the milk component involved in curd formation by acid and record it in Table C-1.

Table C-1 Evaluation of the Effect of Heat and Acid

Event	Milk component involved
Formation of film on milk surface	
Precipitation on bottom of pan	
Browning on bottom of pan	
Curd formation by acid	

Table C-2 Evaluation of the Effect of Heat and Acid

Liquid	pH	Observation of thickening
Milk		
Vinegar		
Milk + 1 t. vinegar		
Milk + 2 t. vinegar		
Milk + 1 T. vinegar		
Milk + 1 T. 1 t. vinegar		
Milk + 1 T. 2 t. vinegar		
Milk + 2 T. vinegar		

D. Preparation of Scalloped Potatoes—Effect of Tannins on Milk Proteins

Objectives
1. To observe and compare the effects of tannins on milk proteins in scalloped potatoes prepared both with milk alone and with a white sauce.
2. To gain experience in preparing a dish using a white sauce.

Basic Recipe for Scalloped Potatoes

Ingredients

2 c. raw, peeled, thinly sliced potatoes
1 c. whole milk
2 T. butter
2 T. flour
½ t. salt
Aerosol vegetable oil cooking spray

Procedure

1. Spray aerosol vegetable oil cooking spray into a 1-qt. casserole dish. Arrange 2 c. of sliced potatoes in layers in the prepared dish.
2. Melt butter in a small saucepan over low heat.
3. Add flour and salt. Blend well. This mixture is called a roux.
4. Gradually stir in the milk. Blend as well as possible.
5. Cook over medium heat until thick while stirring continuously. After sauce thickens, continue cooking for two more minutes.
6. Pour the white sauce over the potatoes.
7. Bake uncovered in a 350 °F oven for 1 to 1 ½ hours until potatoes are tender and surface is browned.
8. Evaluate the appearance, flavor and texture of the finished products and record in Table D-1.

Variations

1. Control – Prepare the basic recipe for scalloped potatoes as above.

2. Whole milk – Arrange potatoes as described above. Instead of preparing the white sauce (steps 2-5), pour 1 c. whole milk over the potatoes and bake according to basic preparation directions.

Table D-1 Effect of Tannins on Milk Proteins

Preparation	Appearance	Flavor	Texture
Baked with white sauce			
Baked with milk			

E. Vanilla Pudding

Objectives

1. To prepare and compare vanilla puddings made with whole milk and with soy milk.
2. To prepare a pudding variation using yogurt.
3. To become familiar with non-dairy milk substitutes.

Basic Recipe for Vanilla Pudding

Ingredients

$^{1}/_{3}$ c. granulated sugar
3 T. cornstarch
$^{1}/_{8}$ t. salt
2 c. whole milk
1 T. butter
1 t. vanilla extract

Procedure

1. Mix sugar, cornstarch, and salt in a saucepan.
2. Gradually stir in 2 c. whole milk.
3. Cook over medium heat, stirring constantly. Bring to a full boil and boil one minute.
4. Remove pudding from heat.
5. Stir in the butter and vanilla.
6. Spoon into serving dishes and chill well.
7. Evaluate the appearance, flavor, and texture and record observations in Table E-1.

Variations

1. Control – Follow the basic recipe for vanilla pudding using whole milk.

2. Soy milk – Follow the basic recipe for vanilla pudding, but substitute 2 c. soy milk for whole milk.

3. Low-fat vanilla yogurt – Follow the basic recipe for vanilla pudding, but substitute 1 c. 2% milk for 2 c. whole milk. When pudding is finished (it will be very thick), stir in an 8-oz. container of low-fat vanilla yogurt.

Table E-1 Comparison of Vanilla Pudding Products

Preparation Method	Appearance	Flavor	Texture
Basic vanilla pudding – whole milk			
Basic vanilla pudding – soy milk			
Basic vanilla pudding – 2% fat milk & low-fat vanilla yogurt			

F. Milk Foams

Objectives

1. To gain experience in the preparation of milk foams.
2. To compare the ease of preparation, stability, and characteristics of various milk foams.
3. To investigate the effect of temperature on the preparation of whipped cream.
4. To prepare butter from whipping cream.

Basic Milk Foam Recipe

Ingredients

½ c. milk product (variations require heavy cream, evaporated milk, or reconstituted NFMS)

Procedure

1. Use products below to prepare milk foam according to the following steps (2-8).
2. Record the temperature of the product used for the foam.
3. Beat the cream with an electric mixer at high speed until the whipped cream begins to thicken. Then lower the speed and continue beating until stiff peaks form. Do not over-beat.
4. Record the time required to beat the foam in Table F-1.
5. Measure and record the volume of the foam using a glass graduated measuring cup.
6. Record the appearance of the whipped foam.
7. Use a funnel liner to line a funnel that is supported in a graduated cylinder (see Figure 6-1). Place the whipped foam into the lined funnel.
8. Record the volume of filtrate in the cylinder at 15-minute intervals for one hour in Table F-1.
9. Record any change in the appearance of the whipped foam in Table F-1.

Variations

1. Heavy Whipping Cream

   a. Chilled equipment, chilled product – Beat chilled heavy whipping cream using a chilled mixing bowl and chilled beaters according to the basic recipe.

   b. Room-temperature equipment, room-temperature product – Beat room-temperature heavy whipping cream using a room-temperature mixing bowl and beaters according to the basic recipe.

   c. Chilled equipment, chilled product, butter – Beat chilled heavy whipping cream using a chilled mixing bowl and chilled beaters until butter separates from the buttermilk. Press the buttermilk from the butter. Measure and record the weight of the butter and the volume of the buttermilk; it is not necessary to drain the butter in a lined funnel.

Figure 6-1: Funnel and graduated cylinder set up to measure syneresis.

2. Evaporated Milk

   a. Chilled equipment, chilled product – Chill evaporated milk in a freezer tray in the freezer until ice crystals form around the edges. Beat chilled evaporated milk using a chilled mixing bowl and chilled beaters according to the basic recipe. Use a high speed throughout beating.

   b. Chilled equipment, chilled product, and lemon juice – Chill evaporated milk in a freezer tray in the freezer until ice crystals form around the edges. Beat chilled evaporated milk using a chilled mixing bowl and chilled beaters according to the basic recipe, but add 1 t. lemon juice to the evaporated milk at the beginning of whipping. Use a high speed throughout beating.

3. Non-Fat Milk Solids (NFMS) – Prepare ½ c. reconstituted NFMS. Beat reconstituted NFMS using a chilled mixing bowl and chilled beaters according to the basic recipe, but add 1 t. lemon juice to the reconstituted NFMS at the beginning of whipping. Beat to a stiff foam. Use a high speed throughout beating.

Table F-1 Milk Foams

Variation	Initial foam volume	Drainage volume				Whipping time (min.)	Description of foam
		15 min.	30 min.	45 min.	60 min.		
1a. Whipped cream (chilled cream and bowl)							
1b. Whipped cream (room temperature)							
1c. Chilled product to butter		X	X	X	X		
2a. Evaporated milk foam (no lemon juice)							
2b. Evaporated milk foam (with lemon juice)							
3. Reconstituted NFMS foam							

Post-Lab Study Questions

Textbook reference: Chapter 10

Discussion Questions

1. What is buttermilk? Acidophilus milk? Evaporated milk? Sweetened condensed milk? Low-sodium milk? Soy milk? Compare the kcalories and fat content of these products.

2. How did the characteristics of the reconstituted nonfat dry milk and the skim milk differ? Why was there a difference?

3. When milk is heated, what milk components are found in the skin formed on the surface?

4. Which milk components form the precipitate on the bottom of the saucepan?

5. What causes the browning on the bottom of the saucepan?

6. What is the pH at which casein forms a curd?

7. What was the pH of fresh whole milk? At what pH did thickening begin? At what pH was curd formation apparent?

Questions for Post-Lab Writing

8. How did the three vanilla puddings made with different milks differ in appearance, flavor, aroma, and mouthfeel?

9. Describe a foam.

10. Compare the whipping times for the various milk foams.

11. What was the purpose of putting the evaporated milk into the freezer before whipping? Of adding lemon juice?

12 How does the temperature of whipping cream influence the quality of the foam? Why?

13. What liquid was left behind after the butter separated from the whipped cream?

14. What advantage is there to using a foam prepared from evaporated milk? From reconstituted NFMS?

15. How does the texture of the different yogurts compare? What causes this difference?

Unit 7 – Cheese

Introduction

Cheeses are used widely in food preparations such as sauces, casseroles, salads, sandwiches, and pies. Cheese is a gel of casein and mineral salts from which most of the whey has been removed. Varying amounts of fat are trapped in the gel. The small amount of whey serum remaining in the cheese may contain dissolved lactose, whey proteins, and water-soluble vitamins and minerals.

Natural cheeses may be classified according to their degree of firmness, which is an indication of moisture content, and by the degree and method of ripening. Processed cheeses, cheese foods, and cheese spreads are a blend of natural cheeses with varying characteristics to which emulsifiers and other optional ingredients have been added. Their flavors are usually more bland than those of natural cheeses.

Upon heating, cheese loses some of its moisture. The cheese proteins coagulate, toughen, and shrink. The emulsion system breaks when the cheese is heated and the fat melts and separates. Processed cheeses blend more readily into sauces than natural cheeses do. Well-ripened cheeses blend more readily than unripened cheeses.

Key Terms

aged cheese – cheese made from the curd or solid portion of curdled milk that has <80% moisture and is held at a defined temperature and moisture for a designated time, with added curd turning or added ingredients to develop desired texture and flavors. Examples include cheddar, aged cheddar, Swiss, mozzarella, and provolone.

cube – to cut into approximately ¼" to ½" squares.

imitation cheese – cheese-like product that contains vegetable oil instead of the milk fat found in natural cheese.

non-aged or **fresh cheese** – highly perishable soft, whitish-colored, mild-tasting cheese that is made from the curd or solid portion of curdled milk, has >80% moisture content, and is not aged. Examples include cottage cheese, cream cheese, and ricotta cheese.

processed cheese – cheese produced by combining different varieties of natural cheese and mixing them with other ingredients (emulsifiers).

tofu – cheese produced from soy milk.

Pre-Lab Questions

Textbook reference: Chapter 11

1. What is cheese?

2. What occurs during the cheese curing and ripening process? How does the length of the curing period impact the flavor and cooking characteristics of cheese?

3. Predict the difference in the cooking quality of cheddar cheese compared to that of fat-free cheddar cheese. Why do you expect these outcomes?

4. Identify the primary protein source in cheddar cheese, primost cheese, and tofu.

5. What is chymosin? Describe chymosin's effect on milk.

6. Explain the differences among cheese food, cheese spread, and processed cheese.

Lab Procedures

A. Cheese Display

Objectives

1. To become familiar with some of the varieties of cheeses available.
2. To compare the appearance, flavor, aroma, and texture of different types of cheeses.
3. To compare a cheese served at refrigerator temperature with the same cheese served at room temperature.

Basic Procedure for Cheese Evaluation

Ingredients

Assortment of cheeses (see variations for Lab 7B)
Crackers
Apples

Procedure

1. Divide each cheese variety in half.
2. Keep half of each cheese variety at room temperature for at least one hour before tasting.
3. Keep the other half of each cheese variety in the refrigerator until ready to serve.
4. Cut selected cheeses into bite-size pieces just before serving and label all cheese varieties.
5. Serve the cheeses with crackers and sliced apples.
6. Compare cheese varieties for appearance, flavor, aroma, and texture. Record observations in Table A-1.
7. Compare the refrigerated cheese with the same cheese served at room temperature. Record observations in Table A-1.

Table A-1 Evaluation of Cheese Products					
Cheese variety	**Serving temperature**	**Appearance**	**Flavor**	**Aroma**	**Texture**
	Room				
	Refrigerator				
	Room				
	Refrigerator				
	Room				
	Refrigerator				
	Room				
	Refrigerator				

B. Effect of Heat on Cheeses

Objectives

1. To observe the effect of heat on natural and processed cheeses.

Basic Procedure to Evaluate the Effect of Heat on Cheese

Ingredients

8 slices of bread (1 for each cheese slice)
1 slice ($^1/_8$" × 1" × 3") of each cheese (variations designated below)

Procedure

1. Use the assigned cheese. Check the label for % fat, % moisture and presence of emulsifier.
2. Line a baking sheet with aluminum foil.
3. For each cheese variety, place a $^1/_8$" × 3" × 3" slice of cheese on a slice of bread.
4. Broil the samples 6-8" below the broiler until the cheeses melt. Note the presence or absence of stringiness.
5. Transfer the baking sheet to a cooling rack. Observe the appearance of the melted cheeses. Note any fat separation or spreading of the samples.
6. When the samples have cooled just slightly, lift some with a spoon or fork to test for stringiness.
7. Record observations in Table B-1.

Variations

1. Mild cheddar cheese
2. Medium cheddar cheese
3. Sharp cheddar cheese
4. Extra sharp cheddar cheese
5. Process cheese
6. Process cheese food
7. Process cheese spread
8. Fat-free American cheese

Table B-1 Effect of Heat on Cheese Products				
Cheese	**Appearance**	**Fat separation**	**Stringiness**	**Composition (% fat, % moisture, emulsifier present)**
Mild cheddar				
Medium cheddar				
Sharp cheddar				
Extra sharp cheddar				
Process cheese				
Process cheese food				
Process cheese spread				
Fat-free cheese				

C. Preparation of Cottage Cheese

Objectives

1. To compare cottage cheeses prepared by enzyme (chymosin) vs. acid (vinegar) coagulation.
2. To observe the formation and separation of curd and whey in cottage cheese production.
3. To compare homemade cottage cheese with a commercial cottage cheese.

C-a. Enzyme Coagulation

Basic Recipe for Enzyme Coagulation

Ingredients

2 c. skim milk
1 rennet tablet
Water

Procedure

1. Heat 2 c. of skim milk to 98.6 °F in a small saucepan.
2. In a small bowl, dissolve a rennet tablet in 1 T. of lukewarm water.
3. Add the dissolved rennet to the warmed milk and let stand undisturbed at room temperature until a firm gel forms.
4. Slice the curd into small cubes, using a vertical and then horizontal slicing pattern.
5. Set the saucepan in a skillet containing warm water and reheat the curd to 98.6 °F. Maintain this temperature until the whey separates from the curd.
6. Drain the mixture using a cheesecloth bag until the dripping stops. Gently squeeze out remaining whey.
7. Collect the whey and measure its volume. Use pH paper to determine its pH. Observe its color when held against a backdrop of white paper.
8. Collect the cottage cheese curds, weigh, and record observations in Table C-1.

C-b. Acid Coagulation

Basic Recipe for Acid Coagulation

Ingredients
2 c. skim milk
2 T. vinegar + extra if needed

Procedure
1. Prepare clabbered (soured) milk by adding 2 T. of vinegar to 2 c. of skim milk in a small saucepan. Stir well.
2. Let stand 10 minutes. If a gel forms, cut into coarse pieces. If a gel does not form, it may be necessary to add more vinegar.
3. Set the saucepan in a skillet of warm water and heat slowly until the milk reaches 98.6 °F. Hold at this temperature until the whey begins to separate from the curd.
4. Drain the mixture using a cheesecloth bag until the dripping stops. Gently squeeze out remaining whey.
5. Collect the whey and measure its volume. Use pH paper to determine its pH. Observe its color when held against a backdrop of white paper.
6. Collect the cottage cheese curds, weigh, and record observations in Table C-1.

C-c. Commercial Cottage Cheese – Perform steps 4-6 above using a 16-ounce container of commercial cottage cheese and record observations in Table C-1.

Table C-1 Evaluation of Cottage Cheese Curds

	Enzyme coagulation	Acid coagulation	Commercial dry curd
Appearance			
Texture			
Flavor			
Weight of curd			
Volume of whey			
pH of whey			

D. Cheese Sauces

Objectives
1. To gain experience in the preparation of cheese sauces.
2. To compare cheese sauces made with process cheese and cheddar cheese.
3. To observe and compare the effects of added acid or alkaline ingredients in cheese sauces.

Basic Recipe for Cheese Sauce

Ingredients
1 ½ c. milk
2 T. butter
2 T. flour
½ t. salt
3 oz. shredded cheese
1 t. baking soda
1 t. cream of tartar

Procedure

1. Melt 2 T. butter in a small saucepan.
2. Blend in 2 T. flour and ½ t. salt.
3. Add 1½ c. milk, blend well, and cook over medium heat until the mixture thickens. Stir continuously.
4. Continue cooking over low heat for 2 minutes, still stirring continuously.
5. Transfer the white sauce to the top of a double boiler. Be sure the water in the bottom of the double boiler is boiling.
6. Add 3 oz. shredded cheese to the white sauce. Stir constantly. Heat until the cheese is well blended and the sauce is homogenous. Record the time required for the cheese to blend in Table D-1.
7. Determine the pH of the cheese sauce with pH paper.
8. Evaluate the sauce for appearance, consistency, and flavor and record observations and the pH in Table D-1.
9. Divide the cheese sauce into thirds. To one portion, add 1 t. baking soda. To a second portion, add 1 t. cream of tartar. If no effect is observed, heat the two sauces with the added acid and alkali slowly at low heat. Observe the effects and record observations in Table D-1.
10. Determine the pH of the cheese sauces with added acid and alkali and record observations in Table D-1.

Variations

1. Shredded natural cheddar cheese – Use 3 oz. shredded cheddar cheese in preparation of the basic cheese sauce recipe.

2. Shredded American process cheese – Use 3 oz. shredded process cheese in preparation of the basic cheese sauce recipe.

Table D-1 Evaluation of Cheese Sauces		
	Natural cheddar	**American processed cheese**
Blending time		
Appearance		
Consistency		
Flavor		
pH (plain cheese sauce)		
pH (added acid)		
pH (added alkali)		
Effect of added acid		
Effect of added alkali		

Additional Food Preparation Exercises

E. Selected Cheese Dishes

E-a. Chicken Cheese Soup

Chicken Cheese Soup

Ingredients

2 whole boneless chicken breasts
Water
¼ c. margarine
½ c. onion, diced
½ c. carrots, grated
½ c. celery, diced
¼ c. flour
1 ½ T. cornstarch
4 c. chicken broth (from boiled chicken)
4 c. milk
1 lb. Ole English cheese (cheese food), cubed
1 t. salt
¼ t. pepper
1 T. dried parsley

Procedure

1. Place chicken in a 2-quart saucepan and add enough water to cover it. Bring chicken to a boil, reduce temperature to simmer, and cook to an internal temperature of 165 °F.
2. Remove chicken and place it in a pan to cool.
3. In a heavy saucepan, melt margarine and sauté onion, carrots, and celery. Remove from the heat. Stir in flour and cornstarch. Make a smooth paste.
4. Gradually add chicken broth and milk. Blend until smooth. Cook mixture at low temperature until bubbly, stirring constantly.
5. Add salt, pepper, parsley, and cubed cheese. Heat on low temperature and stir until cheese is melted.
6. Chop chicken breasts into ¼" cubes. Add chicken to cheese mixture and mix well.
7. Continue to heat on low temperature to 165 °F. Serve.

Adapted from the recipe file of Karen Beathard

E-b. Cheese Ball

Cheese Ball

Ingredients

16 oz. cream cheese
2.5 oz. Buddig beef, chopped
½ bunch green onions, chopped
¼ c. chopped pecans

Procedure

1. Whip cream cheese. Add chopped beef and green onion and mix well.
2. Shape mixture into a ball. Roll mixture in chopped pecans.
3. Wrap cheese ball in plastic wrap and refrigerate until service.

From the recipe file of Nancy Morse

Post-Lab Study Questions

Textbook reference: Chapters 10 and 11

Discussion Questions

1. How did the serving temperature of cheese affect its textural characteristics? Why?

2. What effect does the application of heat have on cheese proteins? Carbohydrates? Fats?

3. When heated under the broiler, which cheeses had the greatest tendency to brown? Why?

4. Which cheeses had the greatest tendency to separate? Why?

5. Which cheeses were stringiest? Why?

Questions for Post-Lab Writing

6. Explain the difference between natural cheese and process cheese.

7. What is "American cheese"?

8. Which cheese blended more quickly in the cheese sauce preparation? Why? Less quickly? Why?

9. Did the sauces differ in texture? Why?

10 Compare the flavors of the sauces.

11. What is rennet?

12. Why is the curd cut in making cottage cheese? What is the liquid that is observed when the curd is cut?

13. How does temperature of the milk affect the action of chymosin? What is the optimum temperature? Why does this seem appropriate?

14. Why does enzyme-coagulated curd have a different texture than acid-coagulated curd?

Unit 8 – Eggs

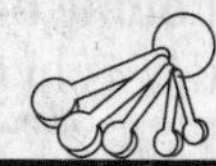

Introduction

Eggs are prepared in a variety of ways. They may be hard- or soft-cooked, poached, baked, scrambled, and fried. They are ingredients in omelets, meringues, custards, and soufflés. Eggs function as thickeners, emulsifiers, binders, and foams in various food products.

Egg yolks contribute flavor, color, fat, and other nutrients to products that contain them. Egg whites provide proteins, which contribute to the structure and stability of products such as foams and custards.

Eggs begin to deteriorate as soon as they are laid. Moisture and carbon dioxide are lost through the pores of the shell. Fresh eggs have a pH of approximately 7.6. Because of carbon dioxide loss, deteriorated eggs may have a pH as high as 9.0 to 9.7. Deterioration can be minimized by keeping the eggs covered and in cold storage.

Egg white foams are stabilized by denaturation of the proteins, which form a protective coating around air bubbles whipped into the liquid egg white. Adding an acid, such as cream of tartar, to an egg white foam decreases the pH of the foam and enhances denaturation of the egg proteins. Thus a more stable foam is produced. Sugar retards denaturation of the egg proteins and, if used, should be added to foam near the end of the whipping process. The sugar also ties up water in the foam, resulting in a more stable foam.

Meringues, which are baked egg white foams containing sugar, can be either soft or hard. Soft meringues are baked at a high temperature (425 °F) for a short time (5 minutes), while hard meringues are baked slowly (30 minutes to one hour) at a low temperature (275 °F).

Custards may be baked to form a gel or stirred to produce a thickened custard, which does not gel. Increasing the concentration of egg in a custard decreases the coagulation temperature of the egg proteins.

Hard-cooked eggs can be cooked in simmering (185 °F) water for 15 minutes. Cooking eggs at a higher temperature or for a prolonged period of time may cause a grey-green discoloration on the egg yolk surface. The discoloration is due to the reaction of hydrogen sulfide, a product of protein denaturation, with iron in the egg yolk. This produces black ferrous sulfide ($H_2S + Fe \rightarrow FeS + H_2$). Increased pH of an egg, as in a deteriorated egg, promotes ferrous sulfide production. The egg white shrinks and becomes tough and rubbery and the yolk also becomes tough and dry.

Key Terms

albumin – egg white, which is comprised primarily of protein and water.

boil – to cook in a liquid that bubbles actively during the time of cooking. The boiling temperature of water at sea level is 212 °F (100 °C).

clarify – to make or become clear or pure. Egg whites are often used to clarify liquids.

coddling – method of egg preparation in which an egg is cracked into a small cup called a coddler. The whole coddler is submerged in simmering water until the egg is cooked.

double boiler – cooking equipment that includes two pans with one pan inserted in the other. The food that is in the top pan is heated by boiling water in the bottom pan.

ice bath – mixture of ice and water used to rapidly cool food or liquid products.

meringue – egg white foam prepared by beating egg whites into foam and adding sugar. The quantity of sugar added determines the softness or hardness of the foam.

percent sag test – objective test used to measure gel strength. Higher percent sag represents a more tender gel.

soft peak stage – stage in foam formation at which peaks fall over as the beater is lifted from the foam.

soufflé – modified omelet that includes a thick base such as a white sauce, an egg white foam, and flavoring ingredients.

stiff peak stage – stage in foam formation at which just the tip of the peak bends over as the beater is lifted from the foam.

sunny side up egg – an egg cooked until the white is set and the yolk is still soft.

Pre-Lab Questions

Textbook reference: Chapter 12

1. What causes the ferrous sulfide ring in a boiled egg? How can this be prevented?

2. How do the characteristics of an egg change as it ages?

3. Why do egg whites make a stable foam when beaten?

4. Identify two problems that may occur when preparing soft meringues and their causes.

5. What is the primary ingredient in an egg substitute?

A. Quality of Raw Eggs

Objectives

1. To become familiar with the quality characteristics of fresh, raw eggs.
2. To compare quality characteristics of fresh and deteriorated raw eggs.

Basic Procedure to Evaluate Egg Quality

Ingredients

1 fresh grade A egg
1 deteriorated grade A egg (held at room temperature for at least one week)

Procedure

1. Break one fresh egg out of the shell onto a flat plate. Be careful not to damage the egg.
2. Keep the shell. Inspect the air cell in the large blunt end of the egg shell.
3. Observe the thick and thin egg white; the height, diameter, and color of the yolk; the position of the yolk in the white; and the chalazae and record observations in Table A-1.
4. Repeat steps 1-3 using a deteriorated egg.

Table A-1 Evaluation of Egg Quality

Egg Type	Consistency of egg white	Height of the yolk	Location of the yolk	Size of the air cell in the shell	Price per dozen
Fresh grade A egg					
Deteriorated grade A egg					Not applicable

B. Hard-Cooked Fresh and Deteriorated Eggs

Objectives

1. To observe the characteristics of hard-cooked fresh and deteriorated eggs.
2. To emphasize the standard method for preparing hard-cooked eggs.
3. To emphasize some factors involved in formation of a ferrous sulfide ring in hard-cooked eggs.

Basic Preparation of Hard-Cooked Egg

Ingredients

1 egg
2 c. water

Procedure

1. Place 2 c. of cool tap water in a small saucepan.
2. Use a spoon to add the egg to saucepan. Over high heat, bring the water to a full boil. Place a lid on the pan and remove it from the heat. Let it sit for 15 minutes.
3. Carefully peel the egg. If the shell is difficult to remove, peel it under cool running water.
4. Cut the egg in half lengthwise. Remove the yolk of one half of the egg and place the cut side down on the plate.
5. Observe the color, aroma, and location of the yolk for the egg. Record observations in Table B-1.

Characteristics of a high-quality hard boiled egg: The white should be firm, but not rubbery. The yolk will be dry without a moist appearance and no grey-green surface on the outside of the yolk.

Variations

1. Control, fresh egg – Follow basic preparation of hard-cooked egg as above.

2. Fresh egg, long cooking time – Place an egg in cool water; bring to a boil on high. When water boils, reduce heat to low, place a lid on the pan, and cook for 60 minutes.

3. Deteriorated egg – Follow basic preparation of hard-cooked egg, except use a deteriorated egg.

4. Deteriorated egg, long cooking time – Place a deteriorated egg in cool water; bring to a boil on high. When water boils, reduce heat to low, place a lid on the pan, and cook for 60 minutes.

Table B-1 Evaluation of Hard-Cooked Eggs

Degree of freshness	Cooking time (minutes)	Color	Aroma	Location of yolk
Fresh	15			
Fresh	60			
Deteriorated	15			
Deteriorated	60			

C. Scrambled Eggs

Objectives
1. To become familiar with standard techniques for preparing scrambled eggs.
2. To compare conventionally scrambled eggs with microwave scrambled eggs and with scrambled eggs prepared using an egg substitute.

Basic Recipe for Scrambled Eggs

Ingredients
1 fresh egg, large
1 T. milk, whole
$^{1}/_{16}$ t. salt
1 t. butter

Procedure
1. Using a fork, gently beat together 1 egg, milk, and salt. Set aside.
2. Melt butter in a small frying pan over low heat.
3. When the pan is hot, add the blended egg.
4. Continue cooking on low heat, stirring slowly. Scrape the egg from the bottom and sides of the pan.
5. Cook until the egg appears just slightly moist and creamy. Do not overcook.
6. Evaluate the egg for appearance, flavor, texture, and tenderness. Record observations in Table C-1.

Characteristics of a high-quality scrambled egg: It should have a moist, creamy appearance, not dry or dull looking.

Variations

1. Control – Follow the basic recipe for scrambled eggs.

2. Egg substitute – Follow the basic recipe for scrambled eggs, except use egg substitute equivalent to one egg.

3. Microwave – Mix egg according to basic recipe in a microwave-safe bowl. Microwave at 50% power for 1 minute. Stir egg mixture and microwave for ½ minute. Repeat previous step until egg mixture appears just slightly undercooked. Remove the scrambled egg from the microwave and allow to stand for a minute before serving.

Table C-1 Cooking Scrambled Eggs				
Preparation technique	**Appearance**	**Flavor**	**Texture**	**Tenderness**
Control – direct heat				
Egg substitute				
Microwave				

D. Comparison of Baked and Stirred Custards

Objectives
1. To gain experience in custard preparation.
2. To compare the methods for preparing baked and stirred custards.
3. To observe the effect of protein concentration on the coagulation temperature of egg proteins in a custard.
4. To become familiar with "percent sag," an objective test used to measure gel strength.

D-a. Stirred Custard

Basic Recipe for Stirred Custard

Ingredients
1 c. milk, whole
1 egg, large
2 T. granulated sugar
¼ t. vanilla

Procedure
1. Combine the ingredients by blending gently with a wire whip.
2. Fill the bottom of a double boiler with water so that the water level is ½" below the bottom of the upper pan.
3. Heat the water to 194 °F and hold it at that temperature.
4. Put the blended ingredients in the top of the double boiler.
5. Heat, stirring continuously. Maintain the water temperature in the bottom of the double boiler at 194 °F. Do not boil the water.
6. Remove custard from heat when it thickens, which is usually about 176 °F. Record the temperature. The custard should have the same consistency as thick whipping cream and should coat the spoon. Add vanilla.
7. Immediately place the top of the double boiler of custard in an ice water bath. Stir to cool the custard to room temperature (~78 °F).
8. Evaluate custard for appearance, aroma, flavor, and consistency and record observations in Table D-1.
9. Determine viscosity by performing a linespread test (see Appendix A) on the room-temperature stirred custard. Record data in Table D-1.

Characteristics of a high-quality stirred custard: light yellow with a medium consistency, mounding softly onto a spoon. It will be mildly sweet with vanilla to cover up the egg flavor.

D-b. Baked Custard

Basic Recipe for Baked Custard

Ingredients
1 c. milk, whole
1 egg, large
2 T. granulated sugar
¼ t. vanilla
Sprinkle of nutmeg

Procedure
1. Preheat oven to 350 °F.
2. Combine the ingredients by blending gently with an electric mixer.
3. Divide the custard equally among three custard cups.
4. Place cups in a square metal pan. Pour boiling water in the pan to the level of the custard in the cups (see Figure 8-1).
5. Bake until the custards are firm in the center or insert a knife into the outer edge. If the blade comes out clean, the custard is set.
6. Cool custard to 80 °F and do a percent sag (see Appendix A) test on two of the baked custards to measure gel strength.
7. Evaluate the third baked custard for appearance, aroma, flavor, and texture. Record all data in Table D-1.

Characteristics of a high-quality baked custard: a smooth, light-yellow and brown surface. It will have enough structure to hold its shape when cooled and inverted. There will be no weeping or syneresis. A bubbly, glossy surface and weeping indicate excessive heat.

Table D-1 Comparison of Baked and Stirred Custards

Custard	Appearance	Aroma	Flavor	Consistency of texture	Test results
Stirred					**Linespread:**
Baked					**% sag:**

Figure 8-1: Setting up a hot water bath.

E. *Factors Influencing Egg White Foams*

Objectives

1. To identify the factors contributing to volume and stability in egg white foams.
2. To gain experience in the preparation of egg white foams.

Basic Recipe for Egg White Foams

Ingredients

1 fresh large egg, room temperature

Procedure

1. Separate egg white and yolk.
2. Beat the egg white at room temperature with an electric mixer at high speed until just the tip of the peak bends over as the beater is lifted from the foam (stiff peak stage). (See Figures 8-2 to 8-4.)
3. Record the time required to beat the foam to the stiff peak stage in Table E-1.
4. Remove ¼ c. of foam and float on ½ inch (1.5 cm) of hot water in a 8 × 8 inch baking pan. Bake at 350 °F until the egg foam turns brown (~15 min.). Cool and record description in Table E-1.
5. Measure the remaining foam and record the volume of the foam using a glass graduated measuring cup.
6. Record the appearance of the whipped foam.
7. Use a funnel liner to line a funnel that is supported in a graduated cylinder (see Figure 8-5). Place the whipped foam into the lined funnel.
8. Record the volume of filtrate in the cylinder at 15-minute intervals for one hour in Table E-1.
9. Record any change in the appearance of the whipped foam.

Characteristics of a high-quality egg foam: shiny white, small air cells, stiff enough to create a peak when the beater is lifted out of the mixture. A small amount of syneresis.

Variations

1. Control – Prepare egg foam according to the basic recipe.

2. Added sugar – Prepare egg foam according to the basic recipe, except add 2 T. granulated sugar gradually at the soft peak stage (peaks fall over as the beater is lifted from the foam). Then complete beating to the stiff peak stage.

3. Added acid – Prepare egg foam according to the basic recipe, except add ¼ t. cream of tartar at the foamy stage (mixture is still fluid and has many bubbles on the surface). Then complete beating to the stiff peak stage.

4. Splenda® – Prepare egg foam according to the basic recipe, except add 2 T. Splenda® gradually at the soft peak stage (peaks fall over as the beater is lifted from the foam). Then complete beating to the stiff peak stage.

5. Added egg yolk – Prepare egg foam according to the basic recipe, except add ½ t. egg yolk to the egg white before beating.

6. Dry stage – Prepare egg foam according to the basic recipe, except over-beat until the foam is dry and appears dull and curdled.

Figure 8-2: Beaten egg whites at the foamy stage.

Table E-1 Comparison of Egg White Foams

Variation	Initial foam volume	Drainage volume				Whipping time (min.)	Description of foam	
		15 min.	30 min.	45 min.	60 min.		Raw	Cooked
Egg white								
Egg white + sugar								
Egg white + cream of tarter								
Egg white + Splenda®								
Egg white + egg yolk								
Egg white, dry stage								

Figure 8-3: Beaten egg whites at the soft peak stage.

Figure 8-4: Egg foams; soft peak stage on the left, stiff peak stage in the middle and on the right.

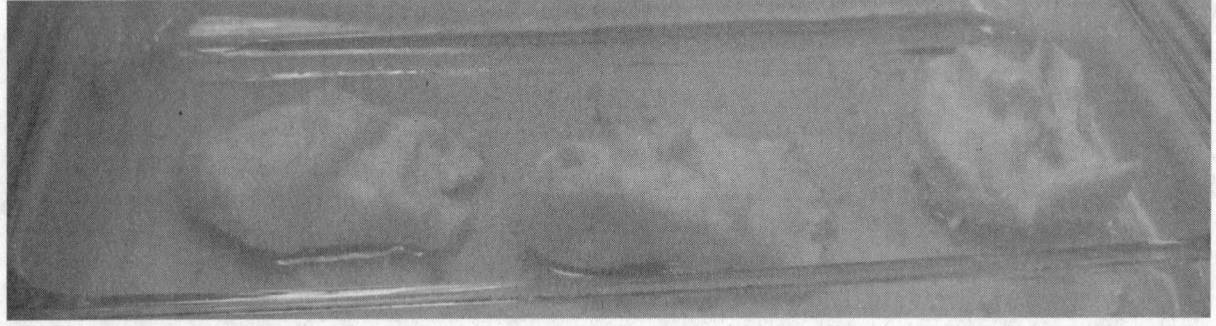

Figure 8-5: Setting up an egg foam in a funnel for measurement of syneresis.

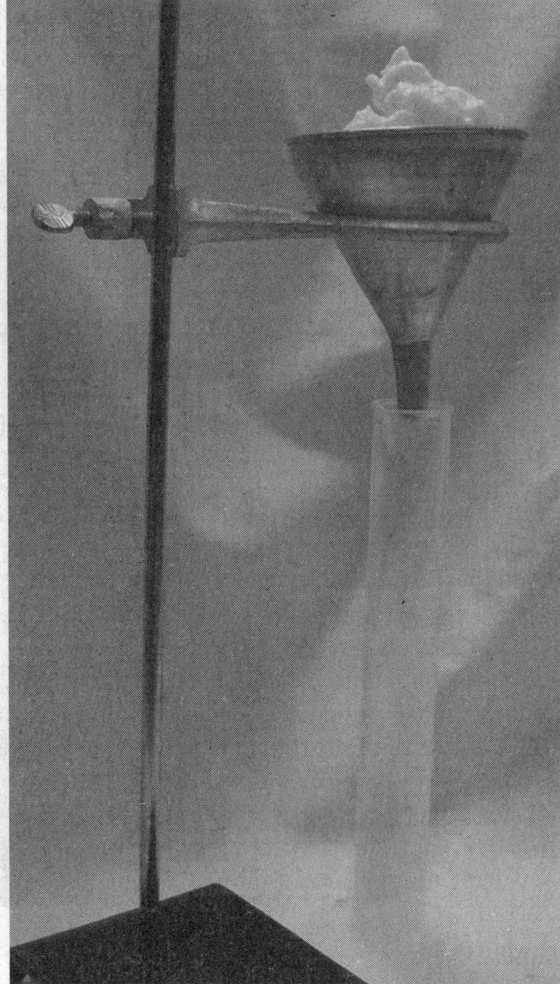

F. Preparation of Angel Food Cake Using Egg Foams

Objectives

1. To demonstrate the preparation of an angel food cake.
2. To apply the use of an egg white foam in angel food cake preparation.
3. To compare angel food cake products prepared with natural egg whites and egg substitutes.

Basic Recipe for Angel Food Cake

Ingredients

1 c. cake flour
1 ½ c. sugar
1 ½ c. room-temperature egg whites (approx. 10 large eggs)
1 ½ t. vanilla extract
1 ½ t. cream of tarter
½ t. salt

Procedure

1. Sift ¾ c. of the sugar and the flour together twice and set aside.
2. Beat the egg whites until foamy.
3. Add the cream of tartar and salt and beat to the soft peak stage.
4. Add the rest of the sugar, a tablespoon at a time, beating thoroughly after each addition.
5. Continue beating to the stiff peak stage.
6. Sift the flour mixture, about one fourth at a time, over the egg whites, folding the mixture in gently with a rubber spatula after each addition using an over-under motion. Avoid over-folding the mixture but be sure all of the dry ingredients have been incorporated.
7. Pour the batter into an ungreased tube pan.
8. Bake at 350 °F for 40 to 50 minutes. Invert the baking pan and allow the cake to cool before removing it from the pan.
9. Evaluate the appearance, aroma, flavor, and texture. Record observations in Table F-1.

Variations

1. Control – Prepare angel food cake according to the basic recipe.

2. Egg white substitute – Prepare angel food cake according to the basic recipe, except substitute 1 ½ c. egg substitute for egg whites. Place the tube pan on a sheet pan when baking in the oven.

Table F-1 Comparison of Angel Food Cake Products

Product	Appearance	Aroma	Flavor	Texture
Angel food cake with egg whites				
Angel food cake with egg substitute				

G. Preparation of Brownies Using Egg Substitutes

Objectives
1. To apply the use of egg substitutes in preparation of brownies.

Basic Recipe for Brownies

Ingredients
1 package brownie mix
Additional ingredients required to prepare brownies from mix (which should include 1 or more eggs)
Aerosol vegetable oil cooking spray

Procedure
1. Prepare brownies according to manufacturer package directions.
2. Perform sensory evaluation on brownies and record comments in Table G-1.

Variations

1. Control – Prepare control brownies according to package directions.

2. Flax seed and water – Prepare brownies according to package directions, except substitute 1 T. milled flax seed and 3 T. water for each egg in the recipe.

3. Soft tofu – Prepare brownies according to package directions, except substitute ¼ c. soft tofu for each egg in the recipe.

4. Water, oil, baking powder – Prepare brownies according to package directions, except substitute 1 ½ T. water, 1 ½ T. oil, and 1 t. baking powder for each egg in the recipe.

5. Egg white substitute – Prepare brownies according to package directions, except replace each egg in the recipe with the equivalent amount of egg substitute.

Table G-1 Evaluation of Brownies

Product	Appearance	Aroma	Flavor	Texture
Control brownies				
Brownies with flax seed and water				
Brownies with tofu				
Brownies with water, oil, and baking powder				
Egg substitute				

Additional Food Preparation Exercises

H. Meringues

Objectives
1. To gain experience in preparing meringues.
2. To compare soft versus hard meringues.

H-a. Soft Meringue

Soft Meringue

Ingredients
1 egg white (room temperature)
2 T. sugar, divided

Procedure
1. Beat egg white with rotary beater until it just reaches the soft peak stage.
2. Gradually add 2 T. sugar and beat slowly until the meringue is very thick but will still form peaks that have rounded tops. Do not beat to the stiff peak stage.
3. Transfer the meringue to a baking sheet and bake at 425 °F for 5-7 minutes until it is lightly browned.

H-b. Hard Meringue

Hard Meringue

Ingredients
1 egg white (room temperature)
4 T. sugar, divided

Procedure
1. Beat egg white with rotary beater until it reaches the soft peak stage.
2. Gradually add 4 T. sugar and continue beating until the foam reaches the stiff peak stage.
3. Transfer the meringue to a baking sheet and bake at 275 °F for 1 hour until the meringue is dry, crisp, and a light brown color.

I. *Selected Egg Dishes*

I-a. Cheese Soufflé

Objectives

1. To become familiar with the preparation of a soufflé.
2. To prepare a food product that contains an egg foam.
3. To observe the physical changes that occur when an egg foam is baked.

Cheese Soufflé

Ingredients

¼ c. butter
¼ t. salt
1 ½ c. milk
¼ c. flour
Dash of black pepper
8 oz. shredded sharp cheddar cheese
2 T chopped, fresh parsley
6 eggs, separated
½ t. cream of tartar

Procedure

1. Preheat oven to 325 °F.
2. In a medium saucepan, melt butter and stir in the salt, pepper, and flour.
3. Slowly stir in the milk. Cook over medium heat, stirring constantly, until the sauce is thick and smooth.
4. Add the cheese. Heat, stirring just until the cheese melts. Remove from heat.
5. In a small bowl, beat the egg yolks slightly.
6. Beat a small amount of the hot cheese sauce into the egg yolks, and then slowly pour the egg yolk-sauce mixture into the remaining sauce, stirring rapidly to prevent lumping. Stir in the parsley. Set aside.
7. In a large bowl, at high speed, beat egg whites until foamy. Then add the cream of tartar and beat just until stiff peaks form.
8. Gently fold the cheese mixture into the beaten egg whites and pour into an ungreased 2-quart casserole or soufflé dish. Bake the soufflé for one hour or until set.

I-b. Ham Quiche

Objectives

1. To become familiar with the preparation of a quiche.

Ham Quiche

Ingredients

1 pastry for 9" pie
1 ½ c. shredded Swiss cheese
1 T. flour
½ c. diced cooked ham
4 eggs, beaten
2 c. half and half
½ t. dry mustard
¼ t. salt

Procedure

1. Preheat oven to 400 °F.
2. Line a 9" pie pan with pastry. Trim excess pastry around edges; fold edges under and flute.
3. Prick bottom and sides of pastry with a fork and bake at 400 °F for three minutes; remove from the oven, and gently prick with a fork. Bake an additional 5 minutes.
4. Reduce oven temperature to 325 °F.
5. Combine cheese and flour; sprinkle evenly into pastry shell. Top with ham. Set aside.
6. Beat eggs and mix with half and half, dry mustard and salt. Pour mixture into pastry shell.
7. Bake at 325 °F for 55 to 60 minutes or until set.
8. Let stand 10 minutes before serving.

I-c. Deviled Eggs

Deviled Eggs

Ingredients

½ dozen eggs
¼ c. salad dressing
1 t. prepared mustard
½ t. salt
Dash pepper
½ T. dill pickle relish
Paprika

Procedure

1. Place 2 c. of water in a small saucepan and heat to 185 °F.
2. Use a spoon to add the eggs to the saucepan and cook for 15 minutes.
3. At the end of the cooking time, cool the eggs under a stream of cold water until they are cool enough to handle. Carefully peel the eggs.
4. Cut the eggs in half lengthwise. Remove the yolks and place in a mixing bowl. Set egg whites aside.
5. Beat egg yolks, salad dressing, mustard, salt, pepper, and pickle relish together thoroughly.
6. Evenly divide the egg yolk mixture and spoon it into the egg white halves.
7. Place deviled eggs on a serving platter. Lightly sprinkle paprika over the top of the deviled eggs. Refrigerate until service.

From the recipe file of Alma Mynhier

Post-Lab Study Questions

Textbook reference: Chapter 12

Discussion Questions

1. How are the depth of the air cell and the age of the egg related?

2. Describe the differences that you observed between the raw fresh and deteriorated eggs.

3. Which hard-cooked egg preparation resulted in the least formation of ferrous sulfide around the yolk? Why?

4. Which scrambled egg preparation technique did you prefer? Why?

5. Are eggs that are over 1 week old satisfactory for scrambling? Why?

6. Compare the percent sag of each baked custard.

7. Compare the linespread value of each stirred custard.

8. Compare the appearance, aroma, and flavor of the custards (stirred vs. baked).

Questions for Post-Lab Writing

9. Compare the coagulation temperatures of the stirred custards.

10. Which ingredients of the custard allow it to form a solid gel?

11. Why is baked custard surrounded by hot water during baking?

12. Which components of egg white are responsible for foam formation?

13. Why should egg white foams be used immediately after they are beaten?

14. How did the addition of cream of tartar affect the stability of the foam? Why?

15. How did the addition of sugar affect the stability of the foam? Why?

16. How did the addition of egg yolk affect the stability of the foam? Why?

17. How did over-beating affect the stability of the foam? Why?

18. When preparing meringues, did either meringue "bead"? Why?

19. Did either meringue "weep" or "leak"? Why?

20. Describe the differences between the hard and soft meringues.

21. When preparing a soufflé, why is it important to beat the egg whites into a foam first and then fold in the yolks and other ingredients gently?

Unit 9 – Vegetables and Fruits

Introduction

The fruit of a plant consists of one or more ripened ovaries together with flower parts that may be associated with the ovaries. The ripened ovary contains the seeds of the plant. Almost any other part of the plant maybe classified as a vegetable. Vegetables include leaves, stems, bulbs, roots, tubers, flowers, sprouts, and seeds. A wide variety of fruits and vegetables are available in the marketplace in fresh, canned, frozen, and dried forms. Fresh fruits and vegetables are often picked before optimum ripeness and are then stored under controlled-atmosphere conditions.

The cut or bruised surfaces of many fruits and vegetables are susceptible to enzymatic oxidative browning. In this browning process, the oxidation of colorless polyphenolic compounds present in fruits and vegetables is catalyzed by enzymes called polyphenolases (polyphenol oxidases). The resulting quinone is rearranged, undergoes oxidation, and is polymerized to produce colored melanin. Enzymatic oxidative browning may be inhibited by lowering the pH, by addition of an antioxidant, by immersion in a dilute sodium chloride solution or a sugar syrup, or by blanching the fruit or vegetable.

Objectives of fruit and vegetable cooking include minimizing vitamin and mineral losses, maximizing the development of desirable textures, and maximizing the retention of characteristic desirable flavor compounds. Cooking results in both texture and flavor changes. The texture becomes less firm as structural tissues soften and pectic substances are hydrolyzed. Starch gelatinization also contributes to softening in starchy vegetables. The texture of cooked fruit is affected by the medium in which the fruit is cooked. If the fruit is heated in water alone, water diffuses into the fruit to equalize the differing solute concentrations. The fruit tissues swell and burst, producing a soft, mushy texture. However, when fruit is heated in a sugar syrup that is more concentrated than the cell sap, water diffuses out of the tissues. The loss of water results in shrinking of the fruit and produces a firm texture. Flavor changes in cooked fruits and vegetables depend on the time of cooking and on the specific cooking technique used. In general, a short cooking time will result in a more flavorful fruit or vegetable and also minimize the loss of nutrients.

Vegetable and Fruit Pigments

The pigments in fruits and vegetables are divided into three classes on the basis of chemical structure and properties. These classes include chlorophyll, the carotenoids, and the flavonoids. Chlorophyll and the carotenoid pigments are fat soluble, while the flavonoid pigments are water soluble. The pH of the cooking medium and the length of cooking time are critical factors in determining the color of fruits and vegetables containing these pigments.

In the presence of heat and acid, the magnesium atom in the green chlorophyll molecule is replaced by two hydrogen atoms to form a molecule of olive-green pheophytin. The enzyme chlorophyllase can hydrolyze the phytol side chain of chlorophyll and produce the bright-green compound called chlorophyllide. Chlorophyllin, an intense green compound, results when the phytol and the methyl ester side chains on the porphyrin ring of the chlorophyll molecule are hydrolyzed. This will occur if sodium bicarbonate (baking soda) is added during cooking of a green vegetable.

Carotenoid pigments range in color from pale yellow to orange-red and are relatively stable at acid or alkaline pH values. Carotenoid pigments include the carotenes and the xanthophylls. Carotene pigments can range from pale yellow to red hues, depending on the arrangement of double bonds in the molecule.

Examples of carotene pigments include beta-carotene in carrots and lycopene in tomatoes. Xanthophyll pigments are yellow in color.

Flavonoid pigments constitute a heterogeneous group of compounds that include anthocyanins and anthoxanthins. Flavonoids are sensitive to changes in pH and to the presence of some metals. Anthocyanins change from red in an acid medium to blue in a more alkaline environment, while anthoxanthins change from white or cream-colored in an acid medium to yellow in a more alkaline environment. Examples of foods containing predominantly anthocyanin pigments are cherries, blueberries, red cabbage, strawberries, plums, and grapes. Some foods that contain predominantly anthoxanthin pigments are potatoes, onions, cauliflower, and turnips.

Key Terms

blanch – to dip food briefly in boiling water to inactivate enzymes or to remove skins from fruits, nuts, and vegetables.

chop – to cut into small pieces with a knife or chopper.

cut – to divide food materials with a knife or scissors. Common cutting styles include slicing, shredding, dicing, mincing, and peeling.

dice – to cut into small, even-size cubes, usually ¼" or smaller.

enzymatic browning – browning that occurs when an enzyme acts on a phenolic compound in the presence of oxygen.

grind – to reduce to particles by cutting, crushing, or grinding.

mince – to cut or chop fine using a knife or chopper.

pare / peel – to remove a skin or outside covering using a knife or mechanical peeler.

shred – to cut or tear into thin strips or pieces. Often a knife or a food processor shredder attachment is used to shred food.

steam – to cook in steam with or without pressure.

toss – to mix lightly; usually refers to salad ingredients.

Pre-Lab Questions

Textbook reference: Chapters 13 and 14

1. What are the two main advantages of a short cooking time for fruits and vegetables?

2. Name 4 ways to minimize enzymatic browning.

3. Name a fruit and a vegetable that contain anthocyanin pigments.

4. How do heat and acid affect the green chlorophyll molecule?

5. Will the use of a lid in cooking influence the pH of the cooking liquid? How?

Lab Procedures

A. Enzymatic Oxidative Browning

Objectives
1. To observe the effects of various treatments on development of browning on the cut surface of apples, potatoes, and lettuce.
2. To explain the mechanism by which various treatments inhibit oxidative browning.
3. To compare browning tendencies of the selected foods.

Basic Procedures to Evaluate Enzymatic Oxidative Browning

Ingredients

Sample of fruit or vegetable
Pineapple juice
Lemon juice
Fruit-Fresh®
Sugar
Ascorbic acid solution (made with Fruit-Fresh®)
Cream of tarter solution
Sucrose solution

Procedure
1. Cut sample into ten uniformly sized pieces, leaving the peel on (see Figure 9-1 for correct cutting technique). Each piece should fit into a custard cup.
2. Put each piece of the sample into a separate custard cup.
3. Leave one piece of the sample exposed to air.
4. Blanch one piece of the sample in boiling water for 1 minute.
5. Cover one piece of the sample with pineapple juice. Determine the pH of the juice with pH paper.
6. Cover one piece of the sample with lemon juice. Determine the pH of the juice with pH paper.
7. Cover one piece of the sample with ascorbic acid solution. Determine the pH of the solution with pH paper.
8. Sprinkle dry ascorbic acid (Fruit-Fresh) on one piece of the sample.
9. Cover one piece of the sample with a cream of tarter solution. Determine the pH of the solution with pH paper.
10. Cover one piece of the sample with a sucrose solution
11. Sprinkle granulated sugar on one piece of the sample.
12. Wrap one piece of the sample, untreated, in plastic wrap.
13. Then cover the samples in the custard cups, store in the refrigerator, and observe again at the end of class. Record all observations in Table A-1.

Figure 9-1: Correct use of a knife and cutting board.

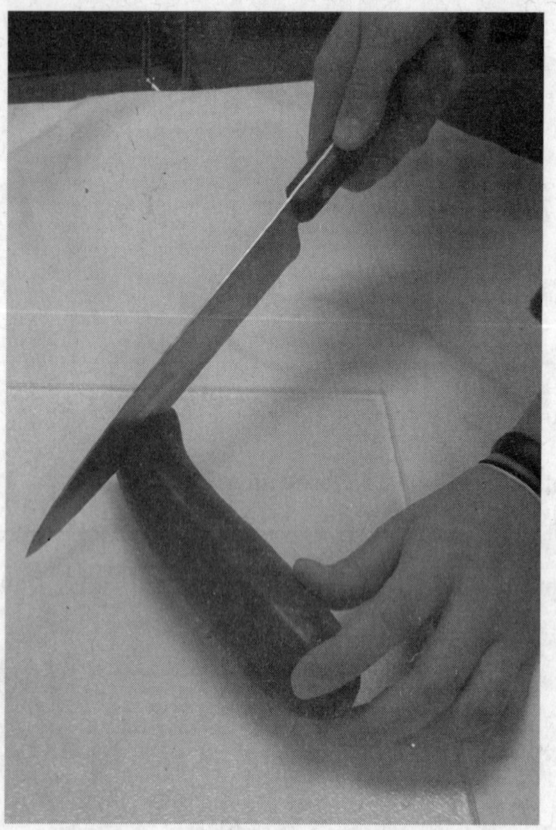

<u>Variations</u>

1. <u>Tart apple (Granny Smith)</u> – Obtain a tart apple (Granny Smith) and follow the basic procedure above. Do not peel the apple.

2. <u>Sweet apple (Red Delicious)</u> – Obtain a sweet apple (Red Delicious) and follow the basic procedure above. Do not peel the apple.

3. <u>Potato</u> – Obtain a potato and follow the basic procedure above. Do not peel the potato.

4. <u>Iceberg lettuce</u> – Obtain iceberg lettuce leaves and follow the basic procedure above.

Table A-1 Enzymatic Oxidative Browning				
	Tart apple	Sweet apple	Potato	Lettuce
Exposed to air				
Blanched				
Pineapple juice **pH =** ______				
Lemon juice **pH =** ______				
Ascorbic acid – solution **pH =** ______				
Ascorbic acid – dry				
Cream of tarter solution				
Sucrose – solution				
Sucrose – dry				
Plastic wrap				

B. Effect of Cooking Medium on Apples

Objectives

1. To observe the effects of cooking apples in varying concentrations of sugar syrup.
2. To observe the effects on texture, appearance, and flavor of fruits cooked in water, sugar syrups, and sugar substitute solution.

Basic Recipe for Cooked Apples

Ingredients

2 apples

Procedure

1. Peel and core two apples. Cut the apples into quarters.
2. Heat 2 c. water on high to boiling in a saucepan. Add the apple quarters. Lower heat to medium-low.
3. Simmer until tender (fork will slide easily into the apple). Keep track of the cooking time required for tenderization to occur, and record in Table B-1.
4. Evaluate appearance, texture, and flavor of apples and record observations in Table B-1.

Variations

1. 2 c. water, no sugar – Prepare apples according to the basic recipe above.

2. 1 c. water, ¼ c. sugar – Follow basic recipe for cooking apples, except reduce water to 1 c. and dissolve ¼ c. sugar in the boiling water before adding the apples.

3. 1 c. water, 1 ¼ c. sugar – Follow basic recipe for cooking apples, except reduce water to 1 c. and dissolve 1 ¼ c. sugar in the boiling water before adding the apples.

4. 1 c. water, 1 ¼ c sugar substitute (Splenda®) – Follow basic recipe for cooking apples, except reduce water to 1 c. and dissolve 1 ¼ c. sugar substitute (Splenda®) in the boiling water before adding the apples.

5. Different apple variety – Prepare a different variety of apples according to each of the previous variations on the basic recipe (1-4). For example, if Delicious apples were used for 1-4, Gala or Jonathan or Winesap apples could be chosen to repeat the procedures. Record cooking time and observations on cooked apples in Table B-2.

Table B-1 Comparison of Effects of Cooking Medium on Apples

Cooking medium	Appearance	Texture	Flavor	Time needed to tenderize
Water				
Water + ¼ c. sugar				
Water + 1 ¼ c. sugar				
Water + 1 ¼ c. sugar substitute				

Table B-2 Comparison of Effects of Cooking Medium on Second Variety of Apples

Cooking medium	Appearance	Texture	Flavor	Time needed to tenderize
Water				
Water + ¼ c. sugar				
Water + 1 ¼ c. sugar				
Water + 1 ¼ c. sugar substitute				

C. *Effects of Cooking Treatments on Vegetable Pigments*

Objectives
1. To observe the effect of cooking time on plant pigments.
2. To observe the effects of acid and alkali on plant pigments.

Basic Recipe for Vegetable Preparation

Ingredients
1 ½ lb. fresh vegetable
2 t. cream of tarter
½ t. baking soda
Water

Procedure
1. Clean fresh vegetables.
2. Divide each vegetable into 6 uniform portions. Use one portion for each cooking method.
3. Use only stainless steel saucepans for cooking vegetables.
4. Cooking methods:
   a. Bring 1 cup water to a boil. Place one portion of vegetables in boiling water and reduce temperature to simmer. Cover after 3 minutes of cooking and cook an additional 7 minutes.
   b. Place one portion of vegetables in a steam basket above 2 cups boiling water. Cover after 3 minutes of cooking and steam an additional 10 minutes.
   c. Bring 1 cup water to a boil. Place one portion of vegetables in boiling water and reduce temperature to simmer. Cover after 3 minutes of cooking and cook an additional 25 minutes.
   d. Add 1 cup water plus 2 t. cream of tartar to a pan and bring it to a boil. Place one portion of vegetables in boiling water and reduce temperature to simmer. Cover after 3 minutes of cooking and cook an additional 7 minutes.
   e. Add 1 cup water plus ½ t. baking soda to a pan and bring it to a boil. Place one portion of vegetables in boiling water and reduce temperature to simmer. Cover after 3 minutes of cooking and cook an additional 7 minutes.
   f. Microwave vegetables on HIGH for 3 minutes in a small covered casserole dish with 1 T. water added.
5. After cooking, drain the liquid into a custard cup and determine the pH with pH paper. Display the cooking liquid. Place vegetables on a white plate for display. Label all samples.
6. Evaluate appearance, texture, and flavor of vegetables and identify the predominant pigment in the vegetable. Note the color of the cooking liquid and of the vegetable.

Characteristics of a high-quality cooked vegetable: a fork will pierce the vegetable with only a little resistance, color should not be diluted or developed into an off-color, flavor will be mild, shape should be intact and not falling apart.

Variations

1. Broccoli – Prepare broccoli according to the basic recipe above. Record observations in Table C-1.

2. Carrots – Prepare carrots according to the basic recipe above. Record observations in Table C-2.

3. Red cabbage – Prepare red cabbage according to the basic recipe above. Record observations in Table C-3.

4. Cauliflower – Prepare cauliflower according to the basic recipe above. Record observations in Table C-4.

Table C-1 Effect of Cooking Treatment on Broccoli

Cooking method	Appearance		pH of liquid	Texture	Flavor	Explanation of observations
	Liquid	Vegetable				
Control						
Steamed						
25 min.						
Cream of tartar						
Baking soda						
Microwave						

Table C-2 Effect of Cooking Treatment on Carrots

Cooking method	Appearance		pH of liquid	Texture	Flavor	Explanation of observations
	Liquid	Vegetable				
Control						
Steamed						
25 min.						
Cream of tartar						
Baking soda						
Microwave						

Table C-3 Effect of Cooking Treatment on Red Cabbage

Cooking method	Appearance		pH of liquid	Texture	Flavor	Explanation of observations
	Liquid	Vegetable				
Steamed						
10 min.						
25 min.						
Microwave						
Soda						
Acid						

Table C-4 Effect of Cooking Treatment on Cauliflower

Cooking method	Appearance		pH of liquid	Texture	Flavor	Explanation of observations
	Liquid	Vegetable				
Control						
Steamed						
25 min.						
Cream of tartar						
Baking soda						
Microwave						

D. Factors That Affect Chlorophyll

Objectives

1. To determine the effect of varying degrees of heat on chlorophyll.
2. To compare the characteristics of fresh, frozen, and canned green beans after cooking.

Basic Recipe for Green Beans

Ingredients

½ lb. fresh green beans
Water

Procedure

1. Wash ½ lb. fresh green beans.
2. Place ½ c. water in a skillet and bring water to a boil.
3. Place fresh green beans into skillet and reduce temperature to simmer. Cover after 3 minutes of cooking and cook until tender. Add more water if needed. Do not let beans cook dry.
4. Once cooked, place green beans on a white plate and pour cooking liquid into a custard cup for display.
5. Evaluate the appearance, texture, and flavor of the green beans. Observe the color of the cooking liquids. Record all observations in Table D-1.

Variations

1. Fresh green beans – Follow basic recipe for green beans above.

2. Canned green beans – Heat a 16-oz. can of green beans. Follow steps 4 and 5 in the basic recipe for green beans.

3. Frozen green beans – Cook a 10-oz. package of frozen green beans according to package directions. Follow steps 4 and 5 in the basic recipe for green beans.

Table D-1 Factors Affecting Chlorophyll

Green bean product	Color		Flavor	Texture	Explanation of color difference
	Liquid	Bean			
Fresh					
Canned					
Frozen					

E. Preparation of Applesauce—Effects of Apple Variety and Cooking Medium

Objectives
1. To observe the change in the texture of applesauce when different varieties of apples are used.
2. To observe the change in applesauce texture when water or a sugar solution is used for preparation.

Basic Recipe for Applesauce

Ingredients
3 medium apples (see variations for variety)
¾ c. water
¼ c. sugar
Sprinkle of cinnamon

Procedure
1. Wash apples. Peel and core; cut each apple into 4 pieces.
2. Place in a saucepan with ¾ cup of water. Heat to simmering. Cover and cook on medium-low for 15 minutes. Check to see if apples are fork tender.
3. Remove from the heat. Mash with a fork until chunky.
4. Blend in sugar.
5. Sprinkle with cinnamon.
6. Evaluate the flavor, texture, and overall quality of the applesauce. Record your observations in Table E-1.

Variations

1. Delicious apples – Prepare basic applesauce recipe using Delicious apples.
2. Jonathan or Winesap apples – Prepare basic applesauce recipe using either Jonathan or Winesap apples.
3. Apples cooked in water + sugar – Prepare basic applesauce recipe using either variety of apples, except add the sugar to the water at the beginning (during step 2), with the apples.

Table E-1 Effects of Apple Variety and Cooking Medium on Applesauce

	Tenderness	Flavor	Overall quality
Delicious			
Jonathan or Winesap			
Water + sugar			

Additional Food Preparation Exercises

F. Fresh Fruits as Ingredients in Food Preparation

Objectives
1. To become familiar with various types of fruits.
2. To demonstrate the use of fruit in food preparation.
3. To evaluate the appearance, texture, and flavor of various fruits.

F-a. Fruit Crumble

Fruit Crumble

Ingredients

2 c. fresh fruit (blueberries, blackberries, raspberries, apples, peaches, pears, or a mixture)
¼ c. sugar (or less, depending on the sweetness of the fruit)
1 T. flour
Pinch of salt
½ t. grated lemon rind
1 ¼ c. vanilla wafer crumbs
¼ c. + 2 T. flour
¼ t. cinnamon
¼ c. melted butter

Procedure

1. Mix the sugar, 1 T. flour, salt. and lemon rind with the fruit.
2. Put the mixture into a casserole dish.
3. Bake at 450 °F for 10 minutes.
4. Reduce the heat to 350 °F and top with a mixture of the crumbs, ¼ c. + 2 T. flour, cinnamon, and butter.
5. Bake at 350 °F for 30 minutes or until fruit is tender and topping is brown and crusty.

F-b. Guacamole

Guacamole

Ingredients

2 ripe avocados, mashed
1 ripe tomato, peeled and finely chopped
1 T. fresh lime or lemon juice
½ t. salt
Tabasco® sauce to taste
2 T. chopped onion
4 T. canned green chilies, finely chopped
Tortilla chips

Procedure

1. Mash avocado pulp well after peeling the avocados and removing the pits.
2. Blend the avocado until smooth.
3. Add the remaining ingredients. Mix well. If a smooth texture is desired, use a blender to mix ingredients.
4. Serve with tortilla chips.

G. Fresh Vegetables as Ingredients in Food Preparation

Objectives

1. To prepare and taste different types of fresh vegetables.
2. To demonstrate various cooking techniques for preparing fresh vegetables.

G-a. Artichoke with Lemon-Butter Sauce

Artichoke with Lemon-Butter Sauce

Ingredients

1 artichoke
½ lemon, sliced thinly
¼ c. butter
2 t. lemon juice
½ t. salt
Water

Procedure

1. Wash the artichoke. Trim the stem and leaf tips. Place the artichoke in a saucepan with salt and water. Enough water should be used to keep the artichoke afloat.
2. Toss lemon slices into the water. Cover the saucepan.
3. Cook 30-45 minutes, until the underside of the artichoke is tender. Test with the tip of a sharp knife.
4. While the artichoke is cooking, prepare sauce by combining melted butter and lemon juice.
5. Drain the artichoke and serve accompanied by the lemon-butter sauce.
6. Dip the stem end of each leaf in sauce and scrape the tender part of the leaf with the teeth.
7. When all the leaves have been eaten, remove the choke to uncover the artichoke bottom. Cut the artichoke bottom into pieces and dip in sauce.

G-b. Braised Carrots and Leeks with Yogurt-Dill Sauce

Braised Carrots and Leeks with Yogurt-Dill Sauce

Ingredients

8 medium carrots, peeled and cut into ¼" slices
1 bunch leeks
2 T. butter
Salt and pepper to taste
1 c. plain yogurt
2 T. fresh dill, chopped (or dried dill to taste)
¼ c. chicken bouillon

Procedure

1. Peel and slice the carrots. Carefully clean the sand from the leeks. Cut the white portions of the leeks into bite-sized pieces. Discard the remainder of the leeks.
2. Sauté carrot slices and leeks in butter for a minute or two over medium heat.
3. Reduce heat to low. Add the bouillon and braise until the vegetables are tender.
4. Season with salt and pepper, if desired.
5. Combine yogurt and dill. Just before serving, stir yogurt-dill mixture into vegetable mixture.

G-c. Spaghetti Squash with Tomato Sauce

Spaghetti Squash with Tomato Sauce

Ingredients

1 spaghetti squash
1 onion, chopped
1 clove garlic, crushed
15 oz. canned stewed tomatoes
12 oz. tomato sauce
1 t. basil
1 t. oregano
2 T. olive oil
1 t. salt
1 ½ t. sugar
Pepper to taste

Procedure

1. Cut squash in half length-wise. Place each half into a large, deep saucepan in 2-3 inches of water.
2. Place over medium-high heat, cover, and steam until tender.
3. While squash is cooking, sauté onions and garlic in oil in a saucepan for 5 minutes.
4. Add the remaining ingredients and simmer until serving time.
5. To serve, scoop out insides of the squash and place in a serving dish. Pour tomato sauce over the squash, stir, and serve warm.

G-d. Steamed Parsnips

Steamed Parsnips

Ingredients

6 fresh parsnips, peeled and thinly sliced
3 T. butter
Salt and pepper

Procedure

1. Slice and peel parsnips.
2. Cook the sliced parsnips in water for 15 minutes or until tender.
3. Drain and add melted butter, if desired.
4. Add salt and pepper to taste.

G-e. Squash Casserole

Squash Casserole

Ingredients

5 fresh yellow squash, sliced
1 c. water
1 medium onion, diced
½ t. salt
Dash pepper
Envelope cornbread mix (ingredients for cornbread)
1 t. sage
1 can (10.5 oz.) cream of chicken soup
1 c. Velveeta© cheese, shredded
Aerosol vegetable oil cooking spray

Procedure

1. Add 1 c. water to saucepan and bring to a boil. Reduce temperature to simmer and add sliced squash, onion, and salt to water. Cover saucepan with a lid and cook squash approximately 7 minutes until tender.
2. Drain squash and set aside.
3. Mix cornbread according to package directions. Add drained squash, sage, pepper and cream of chicken soup.
4. Spray 9" × 13" baking pan with aerosol vegetable oil cooking spray. Pour squash mixture into prepared pan.
5. Sprinkle shredded cheese over squash.
6. Bake at 350 °F for 30 minutes.

From the recipe file of Alma Mynhier

G-f. Glazed Carrots

Glazed Carrots

Ingredients

16 oz. peeled baby carrots
$^1/_3$ c. apricot jam
¼ c. brown sugar
½ t. ground cinnamon
1 T. butter

Procedures

1. Preheat oven to 450 °F.
2. Place foil on a baking sheet and center carrots on foil.
3. Combine apricot jam, brown sugar, and cinnamon and spread over carrots. Top with butter.
4. Bring up foil sides. Double fold top and ends to seal, making one large packet and leaving room for heat circulation. Bake 25 minutes until tender.

From the recipe file of Lesha Emerson

G-g. Broccoli Casserole

Broccoli Casserole

Ingredients

1 26-oz. package frozen broccoli florets
1 c. mayonnaise
1 c. Colby Jack cheese, grated
1 can (10.5 oz) cream of chicken soup
4 oz. cream cheese
2 eggs, lightly beaten
1 sleeve butter crackers, crushed
2 T. butter, melted
Aerosol vegetable oil cooking spray

Procedure

1. Preheat oven to 350 ºF. Spray a 9” × 13” baking dish with aerosol vegetable oil cooking spray.
2. Prepare broccoli according to package directions. Drain broccoli well.
3. Combine drained broccoli, mayonnaise, cheese, soup, and eggs in a large mixing bowl.
4. Mix well and place the mixture in the prepared baking dish.
5. Bake for 35 minutes or until the top is set and browned and bubbling around the edges.

From the recipe file of Alma Mynhier

Post-Lab Study Questions

Textbook reference: Chapters 13 and 14

Discussion Questions

1. Discuss the techniques used to prevent browning and explain the underlying mechanisms. Which technique was most effective?

2. Assuming that solutions of ascorbic acid and vinegar both have the same pH, which would you predict would be more effective in preventing browning? Why?

3. Did both varieties of apples brown to the same extent? Explain.

4. Compare the textural differences observed between apples cooked in water and apples cooked in a sugar solution. Account for these differences. How did the apples cooked with sugar substitute compare?

5. Why were lime juice and tomato added to the guacamole?

6. Describe the texture and flavor of the various types of fresh vegetables and fruits tasted in the laboratory.

7. Compare the colors of the vegetables used when cooked in acid and alkaline cooking water. Which are the predominant pigments in these vegetables?

8. Do any of the pigments change color when heated for a long period of time? Which? Explain the reaction.

Questions for Post-Lab Writing

9. Does microwave cookery cause a greater or a lesser color change than stove-top cooking? Why?

10. Carotenoids are stable to alkali. However, if carrots are cooked in alkaline water, the cooking water is much more yellow than if the carrots are cooked in plain tap water. Give a possible explanation.

11. Describe the texture of vegetables cooked with baking soda. Why is this so?

12. Describe the color of vegetables cooked with baking soda.

13. Compare the textures of the cooked green beans. Explain the differences.

14. Compare the colors of the cooked green beans. Explain the differences.

15. How does the color of red cabbage change at different pHs?

16. Why do certain apple varieties make a better applesauce? What happens to texture when sugar is added to the water early in the cooking process?

Unit 10 – Legumes

Introduction

Legumes include dried beans, peas, and lentils. Although an excellent protein source, legumes are generally deficient in methionine, an essential amino acid. This deficiency can be overcome by combining the legumes with cereal grains, which supply methionine. Cereal grains are generally low in lysine, which is an essential amino acid conveniently supplied by legumes. Cereals and legumes are often called complementary protein sources because their combination provides all of the essential amino acids. Legumes are a staple in the diets of many people throughout the world but are not usually a major food item for families in the United States. Most legumes must be rehydrated before being cooked. Beans may be added to boiling water, boiled for two minutes, removed from the heat, and allowed to soak for one or two hours in the hot water before cooking. Alternatively, dry beans may be soaked overnight in cold water before further cooking. The cooking time of legumes is influenced by various factors, including the conditions under which the beans were stored, the pH of the cooking medium, the mineral salts present in the water used to rehydrate the beans, and the moisture content of the dry legumes. During cooking, legumes increase in volume. One pound of dried legumes, equivalent to 2 ¼ to 2 ½ cups, increases in volume to 4 ½ to 7 cups when cooked.

Key Terms

falafel – spicy chickpeas that are ground, shaped into patties or croquets, and fried.

hummus – paste or dip comprised of mashed chickpeas, tahini, lemon juice, and garlic.

rehydrate – to replace moisture that was removed with drying.

soybeans – seeds of a plant in the legume family that is the rich protein source of tofu.

tahini – thick paste comprised of ground sesame seeds.

Pre-Lab Questions

Textbook reference: Chapter 13

1. What is the purpose of soaking dried legumes?

2. Describe the short-soak method for legumes.

3. Why should legumes not be prepared using hard water?

4. Why should acid (tomato-based products, lemon juice) not be added to legumes until they are well cooked?

5. Explain how legumes and cereals complement each other.

Lab Procedures

A. Preparing Dried Legumes

Objectives

1. To demonstrate how to prepare dried legumes.
2. To observe the effects of various ingredients on the texture, appearance, and flavor of legumes.

Basic Recipe for Dried Legumes

Ingredients

½ c. dried legumes (chosen by instructor)
2 c. water

Procedure

1. Wash beans in colander and remove rocks and/or other debris.
2. Add 2 c. water to beans and soak overnight.
3. Transfer water and beans to saucepan. Add water to cover beans, if necessary.
4. Bring water to a boil. Lower heat to simmer. Cover and simmer beans until tender.
5. Evaluate appearance, texture, and flavor.
6. Use pH papers to determine pH. Record observations, pH, cooking time, and yield in Table A-1.

Variations

1. Control – Prepare dried legumes according to the basic recipe above.

2. Cream of tartar – Follow basic recipe for dried legumes, except add 2 t. cream of tartar at the beginning of cooking.

3. Baking soda – Follow basic recipe for dried legumes, except add ½ t. baking soda at the beginning of cooking.

4. Unsoaked legumes, same cook time – Follow basic recipe for dried legumes, except do not soak legumes overnight. Cook unsoaked legumes the same amount of time as control legumes.

5. Unsoaked legumes, cooked until tender – Follow basic recipe for dried legumes, except do not soak legumes overnight. Cook unsoaked legumes until tender. Record total cooking time and observations in Table A-1.

Table A-1 Evaluation of Prepared Legumes

Cooking method	pH	Yield	Cooking time	Appearance	Texture	Flavor	Explanation of observations
Control							
Cream of tartar							
Baking soda							
Unsoaked legumes, cooked equal time							
Unsoaked legumes, cooked until tender							

Additional Food Preparation Exercises

B. A Variety of Legume Preparations

Objectives

1. To prepare various recipes using legumes.
2. To demonstrate proper techniques for preparing legumes.
3. To emphasize the complementary protein contribution of recipes that combine legumes and cereals.

B-a. Lentil Salad

Lentil Salad

Ingredients

1 c. dried lentils
1 onion stuck with 2 cloves
1 bay leaf
3 c. water
1 t. salt
2 ½ T. oil
1 ½ T. wine vinegar
Finely chopped scallions, to taste
2 T. fresh chopped parsley
Ground black pepper, to taste

Procedure

1. Place the lentils, the onion stuck with cloves, and the bay leaf in a saucepan.
2. Add water and salt and simmer until tender, 30 to 40 minutes.
3. Drain and discard bay leaf and onion.
4. Add the oil, vinegar, and chopped scallions. Cool to room temperature.
5. At serving time, add the parsley and pepper and mix lightly.

B-b. White Beans with Tomatoes and Garlic

White Beans with Tomatoes and Garlic

Ingredients

3 c. canned cannelloni or other white beans
¼ c. olive oil
1 t. garlic, minced
½ t. dried sweet basil
2 large ripe tomatoes, peeled and coarsely chopped
½ t. salt
Dash ground black pepper
1 T. wine vinegar

Procedure

1. Drain canned beans in a large colander, wash under cold running water, and set aside.
2. In a heavy 10-inch skillet, heat the oil until a light haze forms.
3. Add the garlic and basil and cook, stirring, for 30 seconds.
4. Stir in the drained beans, tomatoes, salt, and a dash of pepper.
5. Cover and simmer over low heat for 10 minutes.
6. Taste for seasoning, and stir in the vinegar.

B-c. Herbed Soy Bean Patties

Herbed Soy Bean Patties

Ingredients

16 oz. canned soybeans, drained
⅛ t. garlic powder
½ t. dried thyme
3 T. wheat germ
½ t. salt
Fresh whole-wheat bread crumbs
½ medium onion, chopped finely
¼ to ½ t. rosemary, to taste
½ c. fresh parsley, chopped
1 large egg, beaten
⅛ t. pepper
Aerosol vegetable oil cooking spray

Procedure

1. Mash soybeans until the beans are reduced to very small chunks, but not turned into a paste.
2. Combine beans with remaining ingredients, except bread crumbs.
3. Shape mixture into patties.
4. Coat each patty with bread crumbs. Spray baking dish with aerosol vegetable oil cooking spray. Arrange patties in prepared pan, and bake at 375 °F for 20 to 30 minutes, or until golden brown.

B-d. Chili with Corn

Chili with Corn

Ingredients

3 T. oil
1 onion, finely chopped
1 clove garlic, cut in half
1 green pepper, chopped
2 c. vegetable stock or water
1 c. coarsely chopped, peeled tomato
2 T. tomato paste
1 c. cooked corn
4 c. cooked kidney or pinto beans
½ t. chili powder
¼ t. cumin powder
1 ½ t. salt
1 t. oregano
Ground pepper, to taste

Procedure

1. Sauté onion and garlic clove in oil until onion is soft. Discard garlic.
2. Add green pepper and sauté another 2 to 3 minutes.
3. Add stock, tomatoes, and corn.
4. Mash 2 c. of the kidney beans and add to pot along with remaining whole beans and seasoning. Simmer 30 minutes. If the mixture seems too watery, remove the cover and cook another 10 minutes.

B-e. Falafel in Pita Bread

Falafel in Pita Bread

Ingredients

16 oz. canned garbanzo beans (chickpeas), drained
1 slice whole-wheat bread
2 T. chopped fresh parsley
1 T. tahini (sesame seed paste)
½ t. turmeric
2 eggs, lightly beaten
½ t. salt
$^1/_8$ t. cayenne pepper
Whole-wheat pita bread
Lettuce, tomato, tahini, or yogurt as garnish

Procedure

1. Prepare crumbs from the slice of whole-wheat bread by using a blender.
2. Also using a blender, grind the garbanzos until coarsely pureed. Add the bread crumbs, parsley, tahini, turmeric, eggs, salt, and pepper and mix well.
3. Form the mixture into small patties and fry the patties in a small amount of oil until they are lightly browned on both sides. Drain on paper towels.
4. Serve 2 or 3 patties (falafel) in a whole-wheat pita bread pocket with lettuce and tomato. Offer tahini and plain yogurt on the side to be used as a dressing, if desired.

B-f. Chickpea Appetizer (Hummus)

Chickpea Appetizer (Hummus)

Ingredients

10 oz. (1 ¼ c.) canned chickpeas
1 ½ T. sesame tahini mixed with 1 ½ T. cold water
2 T. lemon juice
½ clove garlic, minced
½ t. salt
¼ t. black pepper

Procedure

1. Rinse chickpeas under cold water and drain.
2. Puree chickpeas in electric blender.
3. Add lemon juice, tahini, garlic, and salt to the pureed chickpeas.
4. Blend until the mixture is smooth.
5. Serve as a dip with crackers, toast, or breads.

B-g. Tofu Fruit Smoothie

Tofu Fruit Smoothie

Ingredients

1 carton tofu
1 c. frozen raspberries, blueberries, or strawberries (or a combination)
2 T. honey (adjust to taste)
½ c. fruit juice

Procedure

1. Combine all ingredients in a blender.
2. Blend until smooth.
3. Serve chilled.

B-h. Soy Milk Fruit Smoothie

Soy Milk Fruit Smoothie

Ingredients

1 c. soy milk (plain or vanilla)
½ medium banana (may be frozen if desired)
2 T. frozen fruit juice concentrate, undiluted
¼ c. frozen fruit such as peaches or berries

Procedure

1. Combine all ingredients in a blender.
2. Blend until smooth.
3. Serve chilled.

B-i. Tofu Broccoli Quiche

Tofu Broccoli Quiche

Ingredients

1 pre-made pie crust, pre-cooked for 12 minutes or until lightly brown
1 lb. broccoli, trimmed and cut into small pieces (or substitute chopped spinach)
1 medium onion, finely chopped
2 cloves garlic, minced
2 T. oil
1 lb. firm tofu, drained
½ c. milk or soy milk
¼ t. Dijon mustard
¾ t. salt
¼ t. ground nutmeg
½ t. ground red pepper
2 T. Parmesan cheese or soy Parmesan cheese

Procedure

1. Cook broccoli until just tender. Drain.
2. Cook onion and garlic in oil in skillet until golden; add broccoli, heat through, and set aside.
3. In blender, puree tofu, milk, spices, and cheese until smooth.
4. Pour into a large bowl and add the vegetable mixture. Combine well.
5. Pour into pre-cooked pie shell and bake at 400 °F for 35-40 minutes or until quiche is set.
6. Allow to sit for 5 minutes before cutting.

B-j. Mocha Mousse

Mocha Mousse

Ingredients

2 ½ c. mashed silken firm tofu
½ c. sugar
$^{1}/_{3}$ c. unsweetened cocoa powder
1 t. orange zest
2 T. freeze-dried instant coffee granules
½ t. ground cinnamon

Procedure

1. Put all ingredients into a food processor or blender.
2. Blend until very smooth and creamy.
3. Spoon into individual dessert dishes if desired.
4. Cover and refrigerate at least one hour before serving.

Post-Lab Study Questions

Textbook reference: Chapter 13

1. What is the impact of the addition of acid to the final texture, appearance, and flavor of the cooked legumes?

2. What is the impact of the addition of baking soda to the final texture, appearance, and flavor of the cooked legumes?

3. Compare the quality of unsoaked beans that are cooked until tender to unsoaked beans that are cooked as long as the control legumes in the basic recipe.

Unit 11 – Cereals and Flours

Introduction

The word "cereal" is derived from the name of the Roman harvest goddess, Ceres. Cereal grains are the dried seeds of cultivated grasses and have been an important part of the human diet for thousands of years. The principal cereal crops include wheat, maize (corn), rice, wild rice, barley, oats, rye, sorghum, millet, and triticale, a high-protein wheat-rye hybrid. Buckwheat, although not a true cereal because it is not a member of the grass family, is normally classified with the cereals because of structural similarities. Cereal grains in some form are produced in every populated area of the world. Each area generally grows the grain best adapted to its soil and climatic conditions. Cereal products are available to the consumer in many different forms. These include flours, meals, starches, ready-to-eat cold breakfast cereals, hot breakfast cereals, and pastas (macaroni, noodles, spaghetti, etc.) as well as the kernels themselves. Cereal products may be whole-grain products, which are made from the entire cereal kernel, or they may be refined products. Refined products are made from cereal kernels that have had the bran and germ removed and are normally enriched with thiamin, riboflavin, niacin, folic acid, and iron. Calcium and vitamin D are optional additives.

Cereals are cooked to increase their digestibility and palatability. The major component of the cereal grain is the complex carbohydrate starch. The cereal starch gelatinizes as it is heated in the presence of water, resulting in an increase in both cereal volume and viscosity. Therefore, the ratio of water to starch is important in cooking cereals and will influence the final product. Cooking also results in a change in the flavor for the cereal, possibly because of the conversion of starch to dextrins and sugars. Although similar cooking techniques apply to most cereals, the proportions of water to cereal and the cooking times vary with specific cereals. Therefore, it is recommended that cooking directions on cereal packages be followed for best results.

The goal in cooking cereal is to prepare a product that is tender, but not sticky, that is free of uncooked lumps of starch, and that has a pleasant flavor. When cereal is milled, grains are fractured and starch granules are exposed on the surfaces of individual grains. If cereals are agitated during cooking, these starch granules are dislodged, thickening the liquid around the starch grains. The result is a gooey consistency, which many find unpalatable. Therefore, stirring should be kept to a minimum when cooking cereal grains. Because most cereals are bland in flavor, they are extremely versatile. They can serve as basic ingredients in variously flavored or spiced recipes and combine well with many other foodstuffs.

Flours

The term *flour* refers to finely ground meal. Using this general definition, "flours" can be prepared from a variety of foodstuffs, including buckwheat, oatmeal, potatoes, rye, rice, soybeans, corn, barley, chickpeas, peanuts, and cottonseed. However, when used in recipes, the term *flour* is understood to mean all-purpose wheat flour, unless otherwise designated. Hard wheat, soft wheat, and durum wheat are milled into wheat flour. Hard wheat is high in protein and is used to produce bread flours (13-14% protein). Soft wheat has a low protein content and is better suited to production of cake and pastry flours (7-8% protein). All-purpose flour (10-11% protein) is generally milled from a blend of hard and soft wheat. Durum wheat is high-protein wheat that is particularly suitable for use in pasta products. Most wheat flours produced are white refined flours, which are milled after the bran and germ have been removed from the cereal kernel. These refined flours are often enriched. Flour also undergoes chemical treatment after milling to hasten the bleaching and maturing of the flour. Whole-wheat flour is milled from kernels that have not had the bran and germ removed. Also available are self-rising flours, to which baking powder and salt have been added, and instant flours, which have been specially processed to have instant blending properties with liquids.

Flour is a major ingredient in all baked products and contributes to their structure. Gliadins and glutenins are classes of proteins found in wheat. When flour and water are kneaded or mixed together, these proteins combine with water to form an elastic protein complex called gluten. The amount of gluten formed in a dough is dependent on the amount of protein in the flour. The gluten forms a continuous network throughout the dough or batter. As leavening gases expand during baking, the gluten strands stretch and the volume of the product increases. The gluten coagulates during baking and gelatinized starch granules become imbedded in the gluten network. This forms the basic structure of baked products. Gluten is normally considered to be a "toughening" agent. Gluten development contributes not only to structure but also to texture and decreases tenderness of baked products. Therefore, high-protein flours are most suitable for breads and low-protein flours are most suitable for cakes and pastries. Gluten development is influenced both by the type of flour used and by the amount of manipulation. It also depends on the quantities of liquid, sugar, and fat in a recipe. Enough water must be available for the desired amount of gluten to form. Sugar competes with proteins for available water and inhibits gluten development. Fats coats flour particles, making it difficult for water to hydrate the gluten-forming proteins, and also inhibits gluten development. Fat is the more effective inhibitor of gluten development.

Key Terms

al dente – refers to the texture of properly cooked pasta, which is tender, yet firm enough to resist the teeth.

bran – outer covering of the cereal grain that protects its endosperm.

endosperm – largest component of the cereal grain that stores the starch and is the basis for flour production.

germ – smallest, most nutrient-dense component of the cereal grain.

kneading – manipulation procedure that includes pushing, stretching, and folding the dough to develop the gluten and create an elastic mass.

Pre-Lab Questions

Textbook reference: Chapters 16 and 17

1. Why do cereals increase in volume and viscosity during cooking? What is this process called?

2. Explain the difference between rolled oats and instant oatmeal.

3. Explain the differences among long-grain white rice, brown rice, converted rice, and instant rice. Compare their nutrient contributions.

4. What is gluten? Which flour is the highest in gluten content? Which flours do not contain gluten?

5. What type of wheat is used in pasta products? Why?

Lab Procedures

A. Market Forms of Oatmeal Cereals

Objectives

1. To compare the cereal : liquid ratios and cooking times required to prepare different market forms of oatmeal.
2. To compare the ease of preparation and palatability of various forms of oatmeal cereals.
3. To determine the effect of using cold water in the preparation of quick-cooking oats.

Basic Recipe for Oatmeal

Ingredients

Oatmeal
Water
Salt

Procedure

1. Prepare one serving of oatmeal cereal according to package directions. Measure the volume of cereal before cooking using a graduated cylinder.
2. Record preparation time of the oatmeal in Table A-1.
3. Measure the volume of the product after cooking.
4. Calculate the % of initial volume according to the following calculation:

$$\% \text{ of initial volume} = \frac{\text{Final volume}}{\text{Initial volume}} \times 100$$

5. Record % of initial volume in Table A-1.
6. Evaluate appearance, flavor, and mouthfeel for all products and record observations in Table A-1.

Variations

1. Regular rolled oats – Prepare 1 serving according to the basic recipe for oatmeal.

2. Quick-cooking oats, boiling water – Prepare 1 serving according to the basic recipe for oatmeal using boiling water.

3. Quick-cooking oats, cold water – Prepare 1 serving according to the basic recipe for oatmeal using cold water instead of boiling water.

4. Instant oatmeal – Prepare 1 serving according to the basic recipe for oatmeal.

Table A-1 Comparison of Oatmeal Cereals								
Market form of oats	Volume			Cereal -to- liquid ratio	Prep. time	Sensory evaluation		
	Before cooking	After cooking	% of initial			Appearance	Flavor	Mouthfeel
Regular rolled								
Quick-cooking (boiling)								
Quick-cooking (cold)								
Instant								

B. Comparison of Types of Rice

Objectives

1. To compare brown rice, long-grain white rice, converted rice, and instant rice with respect to sensory characteristics, preparation time, and volume change during cooking.

Basic Recipe for Rice

Ingredients

Rice
Water

Procedure

1. Prepare one serving of rice according to package directions. Measure the volume of rice before cooking.
2. Record preparation time of the rice in Table B-1.
3. Measure the volume of the product after cooking. Calculate the % of initial volume according to the following calculation:

$$\text{\% of initial volume} = \frac{\text{Final volume}}{\text{Initial volume}} \times 100$$

4. Record % of initial volume in Table B-1.
5. Evaluate appearance, flavor, and mouthfeel for all products and record observations in Table B-1.

Variations

1. Brown rice – Prepare brown rice according to the basic recipe for rice.
2. Long-grain white rice – Prepare long-grain white rice according to the basic recipe for rice.
3. Short-grain white rice – Prepare short-grain white rice according to the basic recipe for rice.
4. Converted rice – Prepare converted rice according to the basic recipe for rice.
5. Instant rice – Prepare instant rice according to the basic recipe for rice.

Table B-1 Comparison of Types of Rice

Rice	Volume			Rice-to-liquid ratio	Prep. time	Sensory evaluation		
	Before cooking	After cooking	% of initial			Appearance	Flavor	Mouthfeel
Brown								
Long-grain white								
Short-grain white								
Converted								
Instant								

C. Comparison of Types of Pasta

Objectives
1. To compare the volume, texture, and flavor of various pasta varieties.
2. To practice the proper preparation of pasta.

Basic Recipe for Pasta

Ingredients

½ c. pasta
2 c. water
½ t. salt
1 T. oil

Procedure

1. Place dry pasta into a graduated cylinder and record its volume in milliliters.
2. Bring water to a full boil on high and add salt and pasta.
3. Reduce heat to medium high and continue boiling pasta for 8 minutes or for the amount of time given on the container. Record preparation time for the pasta in Table C-1.
4. Drain using a colander. Return pasta to sauce pan and toss the oil over pasta to prevent sticking.
5. Return pasta to graduated cylinder and record its volume in milliliters. Calculate the % of initial volume according to the following calculation:

$$\% \text{ of initial volume} = \frac{\text{Final volume}}{\text{Initial volume}} \times 100$$

6. Record % of initial volume in Table C-1.
7. Evaluate appearance, flavor, and mouthfeel for all products and record observations in Table C-1.

Variations

1. Elbow macaroni – Prepare elbow macaroni according to the basic pasta recipe.
2. Egg noodles – Prepare egg noodles according to the basic pasta recipe.
3. Spaghetti – Prepare spaghetti according to the basic pasta recipe.
4. Whole-wheat pasta – Prepare a whole-wheat pasta variety according to package directions.

Table C-1 Comparison of Types of Pasta

Pasta	Volume			Pasta-to-liquid ratio	Prep. time	Sensory evaluation		
	Before cooking	After cooking	% of initial			Appearance	Flavor	Mouthfeel
Elbow	61	121				Robust, Symmetrical	Buttery	Thick, Leafy
Noodles	23?	46				Fluffy	Plain	Bouncy, Springy
Spaghetti	42?	88				Limp	Salty	Slippery
Whole wheat	44?	81				Dark	Bland	Grainy, Dense

D. Preparation of Gluten Ball

Objectives

1. To compare the quantity of gluten in different types of flour.

Basic Recipe for Gluten Ball

Ingredients

1 c. flour

¼ c. water

Procedure

1. Mix 1 c. flour with ¼ c. water until all the water is absorbed.
2. Knead dough 10-15 minutes until it is cohesive and elastic.
3. Rinse the dough using either Method 1 or Method 2, as directed by your instructor:
   a. Method 1: Wrap dough in a large piece of cloth. While holding the cloth-wrapped dough under running water, squeeze it to work out the starch (see Figure 11-1). Continue until the water pressed out is clear.
   b. Method 2: Place dough in a bowl filled with cold water; squeeze the dough to work out the starch. Repeat the process with fresh water until the bowl water is clear.
4. Press water from the dough. Weigh the gluten ball prior to cooking.
5. Bake the dough ball at 400 °F for about 30 minutes or until firm. Weigh the gluten ball.
6. Record data and observations in Table D-1.

Variations

1. Bread flour – Follow the basic recipe for gluten ball using bread flour.
2. All-purpose flour – Follow the basic recipe for gluten ball using all-purpose flour.
3. Cake flour – Follow the basic recipe for gluten ball using cake flour.
4. Whole-wheat flour – Follow the basic recipe for gluten ball using whole-wheat flour.
5. Rye flour – Follow the basic recipe for gluten ball using rye flour.
6. Cornmeal flour – Follow the basic recipe for gluten ball using cornmeal flour.

Table D-1 Comparison of Gluten Content

Type of flour	Precooked weight	Post cooked weight	Volume	Size of cells
Bread flour				
All-purpose flour				
Cake flour				
Whole-wheat flour				
Rye flour				
Cornmeal flour				

Figure 11-1: Procedure for rinsing starch from dough to form a gluten ball.

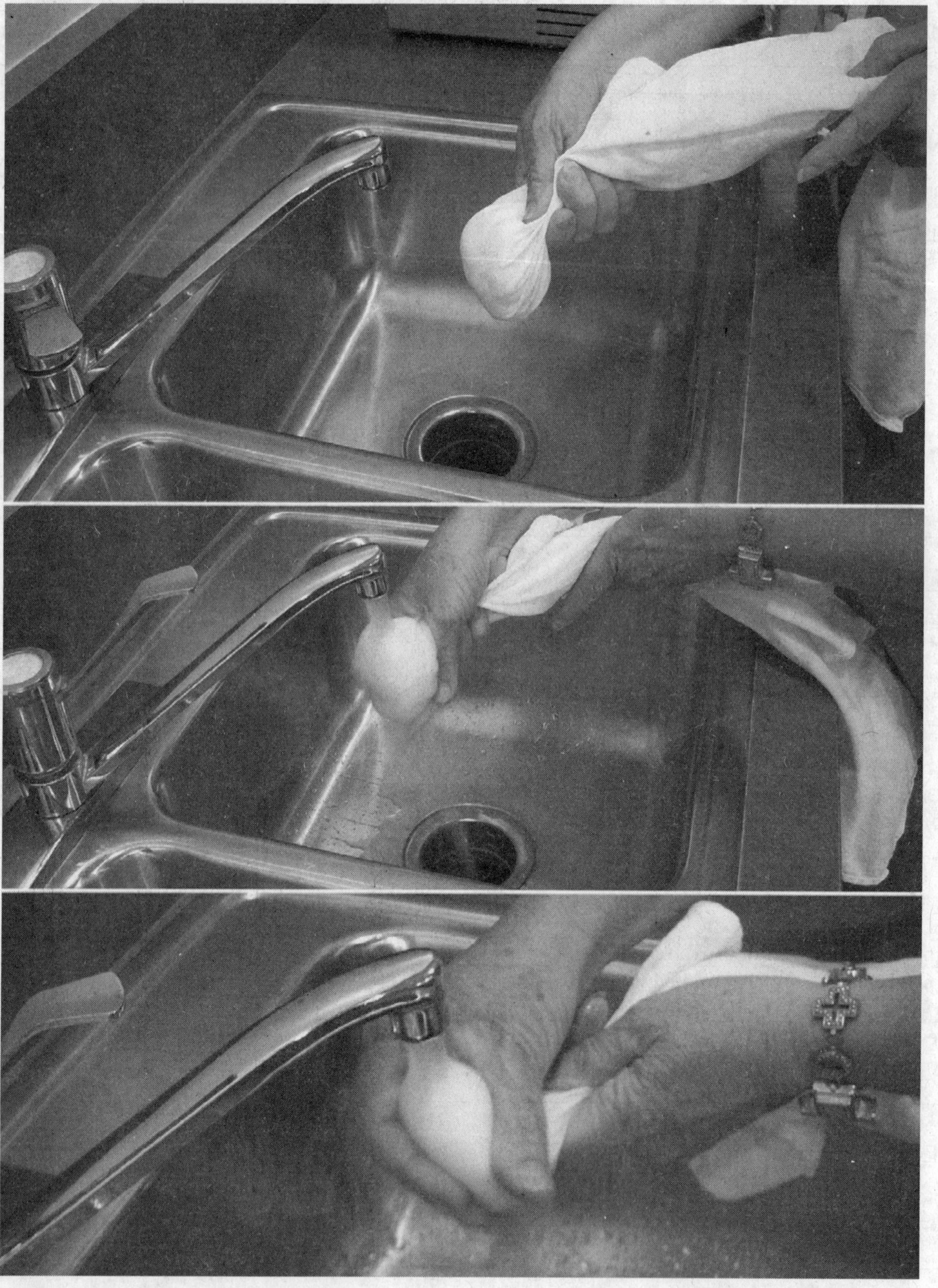

Additional Food Preparation Exercises

E. Cereals as Ingredients in Food Preparation

Objectives
1. To compare the uses of different types of cereal grains.
2. To enhance skills in food preparation.

E-a. Corn

If corn is soaked in an alkaline solution for 20-30 minutes, the kernel softens and the hull is easily removed. The alkaline solution may be prepared by using limestone (calcium hydroxide), lye (sodium hydroxide), or wood ash (potassium hydroxide). After soaking and removal of the hull, the corn is drained and rinsed. If canned at this stage, it is called canned, **whole hominy**. However, it is more often dried and ground after the rinsing procedure. If ground to a coarse consistency, the product is called **hominy grits**. If ground into a flour, the product is called **masa harina,** which is the basic flour for making corn tortillas.

Hominy Grits

Ingredients
Package of hominy grits
Water

Procedure
1. Prepare 1 portion of hominy grits according to package directions.
2. Display some of the dry grits along with the finished product.

Chili Con Queso

Ingredients for Tortillas

2 $^{1}/_{3}$ c. instant masa harina (corn flour)
1 t. salt
1 to 1½ c. cold water

Procedure

1. Combine the masa harina and salt in a deep bowl. Stirring constantly, pour in 1 c. cold water in a slow stream.
2. Knead the mixture vigorously. Add up to ½ c. more water, if necessary, until the dough is firm and does not stick to your fingers.
3. Divide dough into 12 equal portions and roll each one into a ball between your palms.
4. Roll each ball into a 6" circle.
5. Using a moderately hot, lightly oiled skillet, fry each tortilla 1-2 minutes on each side.
6. Keep tortillas warm until served by wrapping them in a towel and placing them in a warm oven.

Chili con Queso Sauce

Ingredients for Sauce

3 T. butter
3 T. flour
8-oz. can tomatoes with jalapeño peppers
4-oz. can green chilies, drained and chopped
6 oz. Monterey Jack cheese, shredded
¼ - ½ c. milk

Procedure

1. Melt butter in a saucepan.
2. Add flour. Mix well and cook over moderate heat for about 1 minute.
3. Add canned tomatoes. Cook over moderate heat, stirring constantly until mixture thickens.
4. Stir in chopped green chilies and shredded cheese. Continue stirring over moderate heat until cheese is melted.
5. If sauce is too thick, add milk until the desired consistency is reached.
6. Keep warm until served by placing in top part of a double boiler.

To Serve

1. Spread a tortilla with chili con queso sauce, roll and eat out of hand or with knife or fork.
2. Alternatively, tortillas may be torn into bite-size portions and dipped into warm sauce.

E-b. Wheat

Wheat is one of the most versatile cereal grains. It is widely used in a variety of ready-to-eat cereals and as a source of flour for baked goods. **Durum wheat**, a special variety of hard, high-protein wheat, is used to produce semolina, the wheat product that is used commercially for all types of pasta. **Farina** is a granular meal made from hard wheat. **Couscous** is also a granular meal made from hard wheat, but the size of the endosperm particles is larger than those in farina. **Bulgur** wheat is whole wheat that has been parboiled, dried, partly debranned, and cracked into coarse angular fragments.

Couscous

Ingredients

1 package couscous
Water
1 T. melted butter

Procedure

1. Prepare ¾ c. couscous as directed on package.
2. To serve, fluff couscous with a fork and toss with 1 T. melted butter.
3. Display some of the dry couscous along with the finished product.

Farina

Ingredients

1 package farina
Water

Procedure

1. Prepare 1 serving of farina according to package directions.
2. Display some of the dry farina along with the finished product.

Almond-Poppy Seed Noodles

Ingredients

6 oz. spinach noodles
¼ c. sliced almonds
1 ½ T. poppy seeds
2-3 T. butter

Procedures

1. Melt butter in a saucepan.
2. Add almonds and sauté until the almonds are lightly browned. Add poppy seeds.
3. Cook the noodles in boiling, salted water until "al dente" according to package directions.
4. Drain and turn into a serving dish.
5. Pour the warm butter-almond-poppy seed mixture over the noodles. Toss and serve.

Tabouleh

Ingredients

2 tomatoes, chopped
1 cucumber, peeled and chopped
½ c. finely chopped fresh parsley
½ c. chopped scallions
2 c. bulgur (cracked wheat)
2 T. olive oil
2 T. lemon juice
1 t. dried mint leaves (optional)
Salt and pepper to taste
Pita bread or romaine lettuce leaves

Procedure

1. Soak bulgur in water while preparing remaining ingredients (about ½ hour).
2. Drain bulgur and squeeze out excess water.
3. Add remaining ingredients and mix well.
4. Serve with pita bread or on romaine lettuce leaves.

Orzo Salad With Tomatoes and Herbs

Ingredients
8 oz. orzo pasta (about 1 ¼ c.)
3 T. sherry wine vinegar
1 T. fresh lemon juice
½ c. extra-virgin olive oil
1 ¼ lb. cherry tomatoes
¾ c. green onions or chives, chopped
½ c. pitted oil-cured olives or pitted Kalamata olives, sliced
¼ c. fresh basil, thinly sliced
¼ c. fresh mint, chopped
¼ c. fresh Italian parsley, chopped
Salt and pepper to taste

Procedure
1. Cook orzo according to package directions.
2. Rinse cooked pasta under cold water and drain well. Transfer pasta to medium bowl and cool.
3. Whisk vinegar and lemon juice in small bowl; gradually whisk in oil.
4. Pour dressing over orzo. Mix in remaining ingredients. Season to taste with salt and pepper.
5. Can be made 2 hours ahead. Let stand at room temperature. Stir before serving.

From the recipe file of Saundra Lorenz - Adapted from a recipe in Bon Appétit *(August 2004)*

E-c. Buckwheat

The major use of buckwheat in the United States is for the manufacture of pancake flour. Silverbull buckwheat is the variety most commonly used for milling purposes because of its high endosperm content. Cooked buckwheat groats, which are called **kasha**, are often served in northern Europe as a side dish.

Buckwheat Pancakes

Ingredients
1 c. buckwheat flour
1 c. all-purpose flour
½ t. salt
2 T. brown sugar
4 t. baking powder
3 T. oil
2 eggs, beaten
1 ¼ - 1 ½ c. milk

Procedure
1. Sift dry ingredients together to ensure complete mixing.
2. In a separate bowl, combine milk, eggs, and oil and mix well.
3. Add liquid ingredients to dry ingredients all at once; stir to combine. Batter may appear lumpy. Do not beat or mix until smooth. Batter should be fairly heavy and not runny.
4. If batter seems too thick, add a small amount of milk and mix well.
5. Cook pancakes on a hot, lightly oiled griddle or fry pan.
6. Cook 1-2 minutes on the second side.
7. Keep warm in oven until served.

Kasha

Ingredients
1 c. kasha (buckwheat groats)
1 egg, beaten
1 t. salt
4 T. butter
½ c. chopped onions
¼ lb. chopped fresh mushrooms
2-2 ½ c. boiling water

Procedures
1. In a mixing bowl, toss kasha and beaten egg together until the grains are coated.
2. Transfer to a skillet and cook uncovered over moderate heat, stirring constantly, until kasha is dry.
3. Add salt, 2 T. butter, and 2 c. boiling water.
4. Stir thoroughly, cover skillet and reduce heat to low.
5. Simmer for 20 minutes. If kasha is still not tender and seems dry, add another ½ cup of boiling water and cook 10 minutes longer until water is absorbed and grains of kasha are tender and fluffy.
6. While kasha is cooking, sauté mushrooms and onions in 2 T. butter, stirring frequently, until liquid in pan has evaporated.
7. Add the mushroom and onion mixture to the cooked kasha and mix lightly.

E-d. Rice

Rice is a staple food for much of the world's population. Long-grain, medium-grain, and short-grain varieties are available. Rice is eaten as a boiled grain and is used in many ready-to-eat breakfast cereals. The whole grain with only the husk removed is called unpolished or **brown rice**. Polished or **white rice** has the husk, bran and germ removed and is usually enriched. **Converted rice** is produced by steeping and steaming rice grains (parboiling) before the bran and germ are removed. This processing technique forces some of the water-soluble vitamins and minerals in the bran and germ to migrate into the endosperm so that when the rice is polished, fewer nutrients are lost. **Quick cooking (minute) rice** has been pre-cooked to gelatinize the starch and subsequently dried. The processed rice has a porous structure that will permit rapid hydration, usually within 5 minutes. **Wild rice** is the seed of a reed-like water plant rather than true rice. The kernels are long, narrow, cylindrical, and dark in color and have a distinctive flavor and texture. The difficulty of its cultivation and harvesting is reflected in its high price. Wild rice is often combined with white or brown rice for the sake of economy.

Rice with Almonds, Onions, and Cheese

Ingredients
1 c. brown rice
2 c. boiling water
½ t. salt
½ onion, chopped
3 T. sliced almonds
2 T. butter
⅓ c. shredded cheddar cheese

Procedures
1. Bring 2 c. water to boiling.
2. Add ½ t. salt and 1 c. brown rice.
3. Reduce heat to low, cover tightly and simmer until rice is tender (40-50 minutes).
4. Meanwhile, sauté chopped onion and almonds in butter until the almonds brown.
5. Just before serving, add onions, almonds, and shredded cheese to the rice and mix lightly.

E-e. Oats

Oats are used primarily as oatmeal. **Oatmeal** is produced by removing the hull from the cereal grain. The germ and the bran are not removed, thereby retaining the minerals, vitamins, and fiber. The oat grains with hulls removed (groats) are passed through rollers to form flakes. **Regular oats** and **quick-cooking oats** differ only in the thinness of the flakes. **Instant oatmeal** has been pre-cooked. Oat flour is also available and may successfully be used in the preparation of baked products when substituted for part of the wheat flour.

Oatmeal Muffins

Ingredients

1 ¼ c. quick-cooking rolled oats
1 ¼ c. milk
1 ¼ c. sifted all-purpose flour
4 t. baking powder
½ t. salt
1 egg, slightly beaten
½ c. oil
⅓ c. firmly packed brown sugar
½ c. raisins (optional)

Procedures

1. Combine oats and milk. Let stand 30 minutes.
2. Combine egg and oil in a separate bowl.
3. Sift flour with baking powder and salt into a large mixing bowl.
4. Mix in brown sugar and raisins.
5. Make a well in the center of the dry ingredients and then add the liquid ingredients all at once.
6. Stir only until all dry particles are moist.
7. Fill 6 well-greased muffin cups two-thirds full.
8. Bake at 400 °F until golden brown: 15-20 minutes for medium muffins; 20 to 25 minutes for large muffins.

Cowboy Cookies

Ingredients

1 c. sugar
1 c. brown sugar
1 c. shortening
2 c. flour
1 t. baking soda
½ t. salt
½ t. baking powder
2 eggs
2 c. oatmeal
1 t. vanilla extract
1 ½ c. chopped pecans
9 oz. chocolate chips
1 ½ c. coconut

Procedure

1. Cream shortening and sugar in a large mixing bowl. Beat in eggs one at a time.
2. Sift flour, baking soda, salt, and baking powder together. Mix flour mixture with shortening mixture.
3. Add vanilla to mixture.
4. Stir in oatmeal, chocolate chips, coconut, and pecans.
5. Drop on greased cookie sheet and bake at 350 ºF for approximately 11 minutes. Yield: approximately 3 dozen cookies.

From the recipe file of Alma Mynhier

Peanut Butter Chocolate Granola Bars

Ingredients

½ c. brown sugar
$^{1}/_{3}$ c. butter
$^{1}/_{3}$ c. crunchy peanut butter
$^{1}/_{3}$ c. honey
1 egg
½ t. vanilla extract
1 c. chocolate chips
3 ½ c. quick oatmeal
¼ c. pecans (optional)
Aerosol vegetable oil cooking spray

Procedure

1. Cream butter and brown sugar.
2. Beat in egg, honey, crunchy peanut butter, and vanilla extract.
3. Stir in oatmeal, chocolate chips, and pecans (if desired).
4. Pour into an 8” × 8” greased pan. Bake at 350 ºF for 25 minutes or until golden brown.

From the recipe file of Karen Beathard

E-f. Barley

Barley is sold mainly in the United States as pearl barley, which is the whole grain with bran and hull removed. It is often used as a soup ingredient. **Barley flour** is used in commercial baby foods and breakfast cereals. **Malt**, used in the manufacture of alcoholic beverages, is a product derived from sprouting barley.

Vegetable Barley Soup

Ingredients

¼ onion, chopped
2 carrots, diced
1 stalk celery, diced
1 turnip, diced
½ c. green beans, sliced
½ c. whole barley, uncooked
2 T. oil
1 qt. hot water
1 t. salt
Pepper to taste
¼ t. marjoram*
¼ t. dried thyme*
1 T. chopped fresh parsley

*Substitute any herbs of your choice.

Procedure

1. Sauté vegetables in oil, 5-10 minutes.
2. Mix in the salt, pepper, marjoram, and thyme.
3. Add the hot water and bring to a boil.
4. Add the barley and again bring to a boil.
5. Cover and simmer for 45 minutes.
6. Sprinkle with fresh parsley before serving.

Barley Pilaf

Ingredients

¾ c. barley, uncooked
2 ½ c. beef broth
⅓ c. chopped onion
¼ c. chopped green pepper
¼ c. sliced celery
1 T. butter

Procedure

1. Bring beef broth to a boil.
2. Stir in barley. Cover, reduce heat, and simmer 30-40 minutes until barley is tender.
3. Meanwhile, sauté onions, green pepper, and celery in butter for about 5 minutes.
4. Stir into cooked barley before serving.

E-g. Gluten-Free Products

Gluten-free diets are required for individuals with celiac disease or sensitivity to gluten. A gluten-free diet eliminates wheat. Traditionally, it also eliminates rye, barley, and oats (due to contamination). This diet modification is difficult as many foods include these grains. Acceptable gluten-free substitutes include rice, soybean, sorghum, and potato flours.

Gluten-Free Banana Muffins

Ingredients

1 c. rice flour
1 c. sweet sorghum flour
1 t. baking soda
½ t. baking powder
¼ t. nutmeg
¾ T. cinnamon
1 egg
1 ½ c. apple juice
2 ripe bananas (mashed)
¼ c. honey
⅓ c. brown sugar
½ c. raisins

Procedure

1. Preheat oven to 350 °F.
2. Combine dry ingredients and slowly add liquid ingredients until the dry ingredients are moist.
3. Line muffin pans with cupcake liners. Fill muffins two-thirds full.
4. Bake 15 minutes or until toothpick inserted in center comes out clean. Muffins will brown quickly.

From the recipe file of Allison Kemp

Gluten-Free Apple Muffins

Ingredients

1 c. rice flour
1 c. sweet sorghum flour
1 t. baking soda
½ t. baking powder
¼ t. nutmeg
¾ T. cinnamon
1 egg
1 ½ c. apple juice
1 medium apple, peeled, cored and diced
¼ c. honey
⅓ c. brown sugar
½ c raisins

Procedure

1. Preheat oven to 350 °F.
2. Combine dry ingredients and slowly add liquid ingredients until the dry ingredients are moist.
3. Line muffin pans with cupcake liners. Fill muffins two-thirds full.
4. Bake 15 minutes or until toothpick inserted in center comes out clean.

From the recipe file of Allison Kemp

Gluten-Free Maple Cookies

Ingredients

1 c. vegetable shortening
½ c. pure maple syrup
½ c. brown sugar
2 eggs
3 c. sweet sorghum flour
½ t. baking soda
1 t. salt
1 ½ t. gluten-free (and corn syrup-free) vanilla
½ t. cinnamon

Procedure

1. Cream shortening, maple syrup, brown sugar, and eggs together. Sift dry ingredients together and add to creamed mixture. Blend in vanilla extract.
2. Transfer cookie dough onto wax paper and shape into a log. Refrigerate overnight.
3. Preheat oven to 350 °F. Slice cookie dough and place on cookie sheets. Bake 8-9 minutes or until golden brown.

From the recipe file of Allison Kemp

Post-Lab Study Questions

Textbook reference: Chapters 16 and 17

1. Why aren't cereals stirred during cooking?

2. Compare the quick-cooking oatmeal products prepared with boiling water and with cold water.

3. What is hominy? What is masa harina? How does masa harina differ from cornstarch?

4. Explain the difference between hard and soft wheat. What is farina? Bulgur wheat?

5. What proportions of water and cereal are used in cooking rice? Why is additional water sometimes used for brown rice?

6. What proportions of water and pasta are used in cooking pasta? How much does the volume of pasta increase with preparation?

7. Compare the texture of the gluten-free muffins to that of the oatmeal muffins.

Unit 12 – Starches and Sauces

Introduction

Starch is a term used to indicate both the amylose and amylopectin molecules and the collection of these molecules organized as granules. Amylose refers to a linear chain of alpha glucose molecules linked by alpha 1—4 glycosidic bonds. Amylopectin refers to a branched chain of alpha glucose molecules linked by both alpha-1,4 and alpha-1,6 glycosidic bonds.

Starch granules are used as thickening agents in soups, sauces, gravies, salad dressings, and desserts such as dessert soufflés, pie fillings, and puddings. Dry starch granules tend to pack together and can be separated by mixing with melted fat, cold liquid, or dry ingredients such as sugar. Separated starch granules absorb liquid more uniformly during the cooking period. Starch is a significant component of cereals and many legumes and accounts for the swelling that takes place in these products as they are cooked.

Gelatinization refers to the changes that occur when a starch is heated in the presence of water. The process of gelatinization is essential to the successful preparation of starch pastes and starch gels. Gelatinization is influenced by the type of starch, the concentration of the starch, and the temperature to which the starch mixture is heated. After gelatinization, some starches will form gels. Gelation is the formation of a three-dimensional starch network. Retrogradation is an extension of the gelation process and is an extensive tightening of the starch network. Retrogradation occurs as the temperature decreases, causing the degree of hydrogen bonding among the starch molecules to increase. Retrogradation is enhanced by refrigeration and freezing of the starch gels. This is irreversible and can create lumps in a thawed sauce.

Different types of starches contribute varying degrees of thickening power and translucency to starch gels. Some starches will exhibit thinning with prolonged heating, and some starches will not form gels. Sugar added to a starch paste decreases viscosity of the paste due to the sugar's ability to tie up water. This results in less water available for swelling of the starch granules. Acid added to a starch paste decreases viscosity of the paste by hydrolyzing the starch to form smaller dextrin molecules. The type of starch used determines the temperature required for gelatinization, the degree of thickening, and the stability of the starch paste on thawing after freezing.

Key Terms

beurre manié – a soft paste thickener prepared by blending equal parts of butter and flour.

blend – to thoroughly mix two or more ingredients together.

dextrinization – process in which starch molecules are broken down into smaller, sweeter-tasting components due to exposure to dry heat.

mother sauces – sauces that serve as a base from which other sauces are prepared. Groups of mother sauces include béchamel (white sauce), espagnole (brown sauce), hollandaise sauce, tomato sauce, and velouté sauce.

modified starch – a chemically or physically modified starch with increased functional characteristics.

retrogradation – seepage of water out of a gel due to continual contraction of the bonds between amylase molecules. This is accelerated by freezing.

roux – a thickener prepared by cooking equal parts of fat and flour.

slurry – a thickener prepared by blending starch and a cool liquid. This combination is then gradually mixed with a simmering liquid sauce base.

Pre-Lab Questions

Textbook reference: Chapter 18; Appendix A

1. Compare amylose and amylopectin, including their thickening abilities.

2. Identify starches that are high in amylose. Identify starches that are high in amylopectin.

3. What is the difference between gelatinization and gelation?

4. What is the expected effect of sugar on the gelatinization of a starch paste?

5. What is the purpose of the percent sag test? (Refer to Appendix A.)

Lab Procedures

A. Vanilla Cornstarch Puddings

Objectives
1. To clarify and demonstrate the process of starch gelatinization.
2. To prepare a vanilla cornstarch-thickened pudding.
3. To compare a homemade pudding, a vanilla pudding mix, an instant vanilla pudding mix, and a high-amylopectin canned vanilla pudding.

Basic Recipe for Homemade Vanilla Pudding (Blancmange)

Ingredients

3 T. cornstarch
¼ c. 2 T. granulated sugar
2 c. milk
$^{1}/_{8}$ t. salt
1 t. vanilla extract

Procedure

1. Mix the cornstarch with the sugar in a saucepan.
2. Add the milk and salt and blend well.
3. Cook over medium-low heat. Stir continuously to prevent the milk from scorching on the bottom of the saucepan and the starch granules from settling.
4. Heat until mixture comes to a full boil. Then boil for 1 minute longer. Remove from the heat.
5. Stir in the vanilla.
6. Pour pudding into individual custard cups. Cover one cup and leave the others uncovered. Cool to 80 °F.
7. Evaluate the appearance, flavor, and texture of the pudding and record in Table A-1. Desirable characteristics: cream colored, glossy top, no lumps, mounds onto spoon, slightly sweet.

Variations

1. Cornstarch, control – Prepare the basic recipe for homemade vanilla pudding as above.
2. Vanilla pudding mix, cooked – Prepare vanilla pudding according to package directions. Pour pudding into individual custard cups. Cover one cup and leave the others uncovered. Chill. Record observations in Table A-1.
3. Vanilla pudding mix, instant – Prepare vanilla pudding according to package directions. Pour pudding into individual custard cups. Cover one cup and leave the others uncovered. Chill. Record observations in Table A-1.
4. Vanilla pudding (high amylopectin), canned – Open vanilla pudding and scoop it into individual custard cups. Cover one cup and leave the others uncovered. Chill. Record observations in Table A-1.
5. Homemade with flour – Prepare the basic recipe for homemade vanilla pudding, except substitute 6 T. flour for the cornstarch.

Table A-1 Comparison of Vanilla Pudding Products

Pudding	Appearance	Texture	Flavor	Type of Starch (see package label)
Cornstarch pudding	Sunken, Smooth white	Firm, ~~springy~~ Smooth	Sweet	
Pudding mix, cooked	Sunken, Dull Dry, rough	springy, Mushy	Lightly sweet	
Pudding mix, instant	Glossy, rounded	Smooth, creamy moist	Vanilla, Sweet	
Canned pudding	Shiny	Creamy, thick Smooth	Very Vanilla	
Flour pudding	Dull, dry	Firm, springy, thick	Floury, mild, Vanilla	

B. *Effect of Starch Variety on Lemon Pie Filling*

Objectives

1. To prepare a starch-thickened pie filling.
2. To compare the effectiveness of various starches on the thickening of a pie filling.
3. To observe the characteristics of pie fillings prepared with various starches.

Basic Recipe for Lemon Pie Filling

Ingredients

2 T. cornstarch
½ c. granulated sugar
¼ c. cold water
¾ c. boiling water
1 egg yolk, beaten
2 ½ T. lemon juice
$^{1}/_{16}$ t. salt
½ t. grated lemon rind
1 T. butter

Procedure

1. Mix cornstarch, salt, and sugar in a 2-qt. saucepan.
2. Blend in the cold water.
3. Add the boiling water and cook over medium heat until thick and translucent, stirring constantly. Mixture should come to a full boil. Then remove from heat.
4. Stir some of the hot mixture into the beaten egg yolk.
5. Return this starch-egg mixture to the remainder of the hot pie filling and stir well.
6. Return to medium heat and stir constantly until thick.
7. Remove from heat and add butter, lemon rind, and lemon juice. Mix well.
8. Pour into custard cups and cool.
9. Evaluate the appearance, flavor, and texture of the pie fillings and record your observations in Table B-1.

Variations

1. Cornstarch – Prepare basic recipe for lemon pie filling as above.

2. All-purpose flour – Follow basic recipe for lemon pie filling, except substitute 4 T. all-purpose flour for cornstarch.

3. Quick-cooking tapioca – Follow basic recipe for lemon pie filling, except substitute 4 T. quick-cooking tapioca for cornstarch.

4. Potato starch – Follow basic recipe for lemon pie filling, except substitute 2 T. potato starch for cornstarch.

5. Arrowroot starch – Follow basic recipe for lemon pie filling, except substitute 3 T. arrowroot starch for cornstarch.

Table B-1 Effect of Starch Variety on Lemon Pie Filling

Starch	Appearance	Flavor	Texture/Mouthfeel
Cornstarch			
All-purpose flour			
Quick-cooking tapioca			
Potato starch			
Arrowroot starch			

C. Effect of Type of Starch on Viscosity of Starch Pastes

Objectives

1. To clarify and demonstrate the process of gelatinization.
2. To compare the behavior and appearance of various starch pastes.
3. To investigate the stability of various starch pastes when thawed after freezing.

Basic Recipe for Starch Pastes

Ingredients

2 T. starch variation
¾ c. water (divided; add gradually to starch)
Ice bath (for cooling)

Procedure

1. Blend assigned starch with ¼ c. cold water in a 1-pt. heavy saucepan to form a smooth paste (see Figure 12-1). Slowly add the remaining ½ c. water and stir well until there are no lumps.
2. Cook at medium heat and stir continuously to prevent lumping. Heat until the paste thickens and reaches a full boil. When full boil is reached, stop stirring and remove from the heat. If the mixture begins to thin before reaching a full boil, immediately stop heating and record the temperature in Table C-1.
3. Let each starch paste cool to 122 °F and do a linespread test (see guidelines and diagram in Appendix A). Record results in Table C-1.
4. Put the starch paste in a custard cup, cover tightly, and cool in an ice bath to 80 °F. Perform a percent sag test (see Appendix A) and a second linespread test on the cooled paste. Record observations in Table C-1.
5. Freeze the remainder of the starch paste in a custard cup, tightly covered.
6. Thaw and unmold the frozen starch pastes during the next laboratory session and compare their visual and textural characteristics. Record observations in Table C-1.

Figure 12-1: Gradually blending a liquid (water) into a starch (all-purpose flour) to create a smooth paste without lumps.

Variations

1. Waxy cornstarch – Prepare basic recipe for starch paste using waxy cornstarch.

2. Cornstarch – Prepare basic recipe for starch paste using cornstarch.

3. Rice starch – Prepare basic recipe for starch paste using rice starch.

4. Quick-cooking tapioca – Prepare basic recipe for starch paste using quick-cooking tapioca.

5. Potato starch – Prepare basic recipe for starch paste using potato starch.

6. Arrowroot starch – Prepare basic recipe for starch paste using arrowroot starch.

7. All-purpose flour – Prepare basic recipe for starch paste using all-purpose flour.

8. Dextrinized flour – Prepare dextrinized (browned) flour by stirring (browning) all-purpose flour in a dry skillet until uniformly brown. Prepare basic recipe for starch paste using dextrinized flour.

9. Whole-wheat flour – Prepare basic recipe for starch paste using whole-wheat flour.

Table C-1 Effect of Type of Starch on Viscosity of Starch Pastes						
Starch	Temp. (°F) – step 2	% Sag	Linespread[1]		Description of Paste	
			122 °F	80 °F	Hot	Cold
Waxy cornstarch						
Cornstarch						
Rice starch						
Quick-cooking tapioca						
Potato starch						
Arrowroot						
All-purpose flour						
Dextrinized flour						
Whole-wheat flour						

1. NOTE: No units of measurement are used.

D. Effect of Acid and Sugar on Viscosity of Starch Pastes

Objectives

1. To demonstrate the effect of added sugar on the characteristics of a starch paste.
2. To demonstrate the effect of added acid on the characteristics of a starch paste.

Basic Recipe for Starch Pastes

Ingredients

Starch variation
Liquid variation

Procedure

1. Blend assigned starch with ¼ c. cold liquid in a 1-qt. heavy saucepan to form a smooth paste. Slowly add the remainder of the liquid and stir well.
2. Cook at medium heat and stir continuously to prevent lumping. Heat until the paste thickens and reaches a full boil. If the mixture begins to thin, immediately stop heating and record the temperature and pH in Table D-1.
3. Let each starch paste cool to 120 °F and do a percent sag and a linespread test (see Appendix A). Record results in Table D-1. Record the pH in Table D-1 if this was not done previously.
4. Put the starch paste in a custard cup, cover tightly, and cool (to 80 °F) in a pan of ice water. Do a second linespread test on the cooled paste. Record observations in Table D-1.

Variations

1. Cornstarch – Prepare basic recipe for starch paste using 2 T. cornstarch and 1 c. water.

2. Cornstarch and sugar – Mix 6 T. sugar with 2 T. cornstarch in a 1-qt. heavy saucepan. Prepare basic recipe for starch paste with cornstarch and sugar combination and 1 c. water.

3. Cornstarch and lemon juice – Mix 4 T. lemon juice with ¾ c. water. Prepare basic recipe for starch paste with 2 T. cornstarch and the lemon juice and water combination.

4. Cornstarch, sugar, and lemon juice – Mix 6 T. sugar with 2 T. cornstarch in a 1-qt. heavy saucepan. Mix 4 T. lemon juice with ¾ c. water. Prepare basic recipe for starch paste with cornstarch and sugar combination and lemon juice and water combination.

Table D-1 Effect of Acid and Sugar on Viscosity of Starch Pastes

Ingredient Variation	Temp. (°F) – step 2	% Sag	Linespread		Description of Paste	pH
			120 °F	80 °F		
Cornstarch						
Cornstarch + sugar						
Cornstarch + lemon juice						
Cornstarch + sugar + lemon juice						

E. Evaluation of White Sauces

Objectives
1. To show the effect of various flour : milk ratios on the consistency of a white sauce.
2. To become familiar with the linespread test.
3. To gain experience in the preparation of a white sauce.

Basic Recipe for White Sauce

Ingredients
2 T. flour
2 T. butter
½ t. salt
1 c. milk

Procedure
1. Melt butter in a small saucepan over low heat.
2. Add 2 T. all-purpose flour and salt. Blend well. This mixture is called a roux.
3. Gradually stir in the milk. Blend as well as possible.
4. Cook over medium heat until thick while stirring continuously. After sauce thickens, continue cooking for two more minutes.
5. Use ¼ c. of the sauce to do a linespread test (see Appendix A). Perform a second linespread test when the sauce cools to 122 °F. Record the outcomes in Table E-1.
6. Observe the consistency and thickness of the sauce and record in Table E-1.

Variations

1. Control – Prepare white sauce according to basic recipe.
2. 1 T. Flour – Prepare white sauce according to basic recipe, except substitute 1 T. flour for 2 T. flour.
3. 3 T. Flour – Prepare white sauce according to basic recipe, except substitute 3 T. flour for 2 T. flour.
4. 4 T. Flour – Prepare white sauce according to basic recipe, except substitute 4 T. flour for 2 T. flour.

Table E-1 Evaluation of White Sauces

Amt. of flour	Linespread		Consistency	Thickness	Suggested use for sauce
	Immediate	122 °F			
1 T.					
2 T.					
3 T.					
4 T.					

Additional Food Preparation Exercises

F. Banana Pudding

Banana Pudding*

Ingredients

Large (5.25 oz.) package instant vanilla pudding
2 c. milk
½ t. vanilla extract
$^2/_3$ c. sugar
8 oz. sour cream
8 oz. Cool Whip
Vanilla wafers
4-6 large bananas

Procedure

1. Whip milk, instant pudding, and vanilla extract until thick.
2. Mix in sugar and sour cream.
3. Fold in 4 oz. Cool Whip.
4. Slice bananas in preparation for layering pudding.
5. Layer vanilla wafers, sliced bananas, and pudding in a 9” × 9” pyrex dish. Repeat layer in same sequence. Top with remaining Cool Whip.

*May reduce fat and caloric content by substituting equivalent amounts of sugar-free and fat-free instant vanilla pudding, skim milk, Splenda®, fat-free sour cream, low-fat Cool Whip, and reduced-fat vanilla wafers for regular vanilla pudding, whole milk, sugar, sour cream, Cool Whip, and vanilla wafers.

From the recipe file of Karen Beathard

Post-Lab Study Questions

Textbook reference: Chapter 18

Discussion Questions

1. Why is continuous stirring necessary when preparing white sauce?

2. What changes took place in the white sauces to indicate that gelatinization had occurred?

3. How did cooling each sauce uncovered affect its appearance?

4. How did the consistencies of the white sauces compare?

5. Why do white sauces set as they cool?

6. Why are starch-thickened sauces, gravies, and puddings cooked further after the mixture has thickened?

Questions for Post-Lab Writing

7. Based on the linespread test, rank the starch pastes in the order from most viscous to least viscous at 122 °F and at 80 °F.

8. Which starch pastes formed a gel at 32 °F (0 °C)?

9. Which starch pastes were clear? Opaque? Stringy or cohesive?

10. Why was there a difference in the appearance of the flour-thickened paste and the cornstarch-thickened paste?

11. Compare the pastes made with all-purpose and whole-wheat flours. Describe any differences.

12. Compare the pastes made with all-purpose and dextrinized flours. Explain any differences.

13. Compare the characteristics of the frozen pastes.

14. What starch would you use if you were preparing a cherry pie that was going to be frozen? Give reasons for your choice.

15. Compare the characteristics of the lemon pie fillings prepared with different types of starch. Which do you prefer?

16. What effect does acid have on the gelatinization of a starch paste? Why?

17. Why was the lemon juice used to flavor the pie filling added at the end of cooking?

Unit 13 – Quick Breads

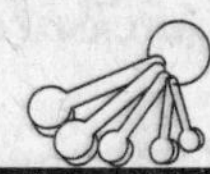

Introduction

Leavening Agents

Leavening means to make light or to rise. A porous, light texture is a desirable characteristic of many baked products. Leavening agents contribute to the development of such a texture by providing a source of gas. Gas bubbles, if produced in an elastic dough that is capable of retaining them and if expanded by the heat of the oven, will cause the dough or batter to increase in volume until the structural framework of the baked product becomes rigid.

The three basic leavening gases commonly used in baked products are air, steam, and carbon dioxide. All baked goods are leavened in part by air that is incorporated during dough preparation and in part by steam because all doughs contain some liquid. However, not all baked products are leavened by carbon dioxide. Air is normally incorporated during creaming of fat and sugar and during mixing of the batter or the dough. Egg white foams that are folded into a batter, as in an angel food cake, also incorporate air. The air cells introduced form pockets. Steam and carbon dioxide from other leavening agents can diffuse into these pockets. Air pockets are essential to good leavening. Steam-leavened products, such as popovers and cream puffs, have a high liquid-to-flour ratio. These products require a hot oven temperature during the first minutes of baking so that water is rapidly converted to steam before the structure of the product becomes rigid. After maximum volume is reached, the oven temperature is lowered so that baking can be completed without over-browning the product. Carbon dioxide is the principal means of leavening flour mixtures. It can be produced biologically by yeast or by specific bacteria such as those used for sourdough breads. It can also be produced chemically by baking powders and by the reaction of baking soda with an acid.

Baking soda can be used in combination with acidic ingredients such as sour milk, buttermilk, yogurt, fruit juices, molasses, brown sugar, and honey. Carbon dioxide is liberated as soon as the baking soda and the acidic ingredient are mixed. Baking powders are commercial mixtures of baking soda and substances that will produce an acid when in solution. Baking powders may be single-acting or double-acting. This is determined by how quickly the acid-reacting substances go into solution. Tartrate baking powder is an example of a single-acting baking powder. It contains tartaric acid and potassium acid tartrate, which are both soluble in cold water. Therefore, carbon dioxide will be released in the dough before the dough is heated in the oven. An example of a double-acting baking powder is SAS-phosphate baking powder. The acid-reacting substances are monocalcium phosphate monohydrate, which is soluble in cold water, and sodium aluminum sulfate, which is soluble in hot water. Therefore only some of the carbon dioxide is produced in the cold dough and the remainder is produced in the dough during baking because the sodium aluminum sulfate does not dissolve and react until the dough is heated. Because the leavening gas is produced in two steps, this is a double-acting baking powder. Most commercial baking powders are double-acting.

Too little leavening agent in a baked product may result in low volume and compact structure. However, too much leavening agent may result in over-expansion of the structure and an open, coarse texture. A greatly over-expanded structure may be weakened and collapse before it sets.

Quick Breads

The term *quick breads* is commonly used to distinguish certain baked products from yeast breads. This group of relatively quickly prepared products is leavened primarily by chemical agents, steam, or air. It includes muffins, biscuits, popovers, griddle cakes, waffles, fritters, dumplings, and a variety of coffee cakes and nut or fruit breads. Quick breads are classified by the proportion of liquid to flour. A pour batter has a liquid-to-flour ratio of 1:1. Popovers and cream puffs are examples of pour batters. Such products are leavened primarily by steam. The egg proteins make a critical contribution to structure because gluten development is minimized. A drop batter has a liquid-to-flour ratio of 1:2 and therefore has a thicker consistency. Muffins are an example of a drop batter. The formation of tunnels in muffins is caused by excess gluten development in the dough during mixing. Soft doughs, such as baking powder biscuits, have a liquid-to-flour ratio of 1:3. Flakiness in biscuits is achieved most effectively by cutting a plastic fat into the dry ingredients. Milk is then added and the mixture stirred until soft dough is formed. Many changes take place in the batter or dough during the baking of quick breads. The fat melts, water-soluble substances, such as salt, baking powder, and sugar dissolve, steam forms and helps to leaven the product, gluten and egg proteins coagulate and contribute to a rigid structure, starch gelatinizes, and both caramelization and carbonyl-amino (Maillard) browning occur.

Key Terms

crumb – interior cell structure of a baked product that is revealed when it is sliced. Cell size, shape, and thickness are criteria for evaluation.

cut in – to distribute solid fat in dry ingredients with knives, with a pastry blender, or in a mixer with proper attachments.

fold in – to gently incorporate one ingredient with another by hand. A large spoon or spatula is used to vertically cut through the mixture and turn it over by sliding the implement across the bottom of the mixing bowl.

leavening agents – physical, biological, or chemical agents that cause a flour mixture to rise.

Maillard reaction – the reaction between a reducing sugar and protein resulting in the browning of foods such as the golden crust of baked products.

stir – to combine two or more ingredients so that they are evenly distributed, using a spoon or other utensil with a circular motion. The spoon is placed at the 12 o'clock position and moved around the bowl.

stroke – when stirring, one complete revolution around the bowl with the spoon (from 12 o'clock to 12 o'clock).

Pre-Lab Questions

Textbook reference: Chapters 17 and 19

1. Explain the differences among baking soda, baking powder, self-rising flour, and yeast and the requirements for leavening activity.

2. Identify leaveners used in quick breads.

3. What are the functions of the high proportions of liquid and eggs in popover batter?

4. Describe the muffin mix method that is most often used in the preparation of quick breads.

5. Why are quick breads mixed only until all ingredients are just moistened?

Lab Procedures

A. Chemical Leaveners

Objectives
1. To determine the volume of leavening achieved with various chemical leaveners.
2. To determine the effect of temperature on release of carbon dioxide and the resulting leavening in a quick bread.

Basic Recipe for Muffins

Ingredients
1 c. flour
¼ t. salt
½ c. whole milk
1 ½ large eggs (~1 ½ T.), beaten
2 T. oil
1 T. sugar

Procedure
1. Preheat one oven to 400 °F. A second oven should be kept cool (not preheated). Grease two muffin tins.
2. Mix dry ingredients together. In a separate bowl, mix liquid ingredients together.
3. Pour liquid ingredients into dry ingredients and stir or mix until ingredients are moist (~ 8 strokes). (See Figure 13-1.)
4. Divide the batter in half, pouring half into each tin. Fill compartments about two-thirds full (see Figure 13-2).
5. Place one tin in a cool oven and set oven to 400 °F (no preheating). Record the time needed for the muffins to become golden brown in Table A-1.
6. Bake the second tin in the preheated oven for 20 minutes, until muffins are golden brown.
7. Cool on a cooling rack for 10 minutes. Place one whole muffin and one muffin that has been cut down the middle, top to bottom, with a sharp serrated knife out for display.
8. Use a ruler to measure the height of the dissected muffin (see Figure 13-3). Evaluate the interior and exterior appearance and flavor of the muffins. Record observations in Table A-1.

Variations

1. No leavener – Prepare the basic recipe for muffins. This variation has no leavening.

2. Soda, ½ t. – Prepare the basic recipe for muffins, adding ½ t. soda.

3. Soda + buttermilk – Prepare the basic recipe for muffins, adding ½ t. soda; use buttermilk instead of whole milk.

4. Baking powder, ½ T. – Prepare the basic recipe for muffins, adding ½ T. double-acting baking powder.

5. Baking powder + standing time – Prepare the basic recipe for muffins, adding ½ T. baking powder. Pour all the batter into a single muffin tin. Allow the batter to sit for 20 minutes before placing it in the preheated oven.

6. Soda + buttermilk + standing time – Prepare the muffin batter as directed in variation #3, but pour all the batter into a single muffin tin. Allow the batter to sit for 20 minutes before placing it in a preheated oven.

Table A-1 Leavening of Muffins

Variation	Browning time	Height	Interior appearance	Flavor	Exterior appearance
1. No leavener, cool oven					
No leavener, warm oven					
2. Soda, cool oven					
Soda, warm oven					
3. Soda + buttermilk, cool oven					
Soda + buttermilk, warm oven					
4. Baking powder, cool oven					
Baking powder, warm oven					
5. Baking powder + standing					
6. Soda + buttermilk + standing					

Figure 13-1: Muffin batter, correctly stirred; note lumpy surface.

Figure 13-2: Proper portioning of muffin batter.

Figure 13-3: Using a ruler to measure muffin height, which relates directly to muffin volume.

B. Leavening by Air

Objectives
1. To demonstrate the use of air as a primary leavening agent in angel food cake.
2. To illustrate the effect of baking conditions on product volume and texture.
3. To emphasize proper mixing techniques for products in which air is the primary leavening agent.

Basic Recipe for Angel Food Cake

Ingredients

1 c. cake flour
1 ½ c. sugar
1 ½ c. egg whites
¼ t. salt
1 ¼ t. cream of tartar
1 t. vanilla

Procedure
1. Sift the flour with ¼ c. of the sugar. Set aside.
2. Beat egg whites to the foamy stage, using a rotary beater or a wire whip.
3. Add the salt and the cream of tartar and continue beating until the soft peak stage is reached.
4. Sprinkle the rest of the sugar over the foam, 2 T. at a time. Beat well after each addition.
5. Beat until the peaks of the meringue just bend over (slightly rounded tops). Fold in the vanilla.
6. Sift ¼ of the flour-sugar mixture over the meringue. Gently fold in the mixture with a rubber spatula, using a down-under-up-over folding motion. Use the minimum number of strokes possible.
7. Repeat Step 6 three more times so that all of the flour-sugar mixture is folded in. No sugar-flour mixture should be visible in the batter.
8. Divide the batter into three portions.
9. Put $^1/_3$ of the batter into an ungreased 9 ¼" × 5 ¼" × 2 ¾" loaf pan and bake immediately in a preheated 425 °F oven for 25-30 minutes or until done.
10. Put $^1/_3$ of the batter into an ungreased loaf pan and bake immediately in a preheated 325 °F oven for 25-30 minutes or until done.
11. Beat $^1/_3$ of the batter for 3 minutes with an electric mixer at high speed. Put the batter into an ungreased loaf pan and bake in a preheated 425 °F oven for 20-25 minutes or until done.
12. Invert pans onto a rack to cool. When cool, remove cake from pans.
13. Evaluate the appearance, volume, and texture of the baked products and record observations in Table B-1.

Table B-1 Leavening By Air

Variation	Appearance	Volume	Texture
425°F			
325°F			
Beat extra 3 min., 425°F			

C. Effect of Egg Protein on Popovers

Objectives

1. To emphasize the role of egg protein in the development of popover structure.
2. To demonstrate the preparation of a steam-leavened quick bread.

Basic Recipe for Popovers

Ingredients

1 c. flour
1 c. whole milk
2 eggs
¼ t. salt
1 T. melted butter
Aerosol vegetable oil cooking spray

Procedure

1. Blend 2 eggs, the milk, and the melted butter with electric mixer.
2. Add flour and salt and beat until the mixture is smooth.
3. Lightly spray muffin tins with vegetable oil cooking spray and fill muffin tins not more than half full.
4. Bake at 450 °F for 15 minutes, then reduce heat and bake 30 minutes at 325 °F.
5. Evaluate the appearance, texture and flavor of the popovers and record observations in Table C-1.

Variations

1. 2 eggs – Prepare basic recipe for popovers as above.
2. 1 egg – Prepare basic recipe for popovers, except use 1 egg instead of 2 eggs.

Table C-1 Effect of Egg Protein on Popovers

Popover	Appearance	Texture	Flavor
2 eggs			
1 egg			

D. Effect of Manipulation and Flour Type on Muffins

Objectives
1. To demonstrate gluten development in muffins.
2. To compare the use of all-purpose, cake, and whole-wheat flours in muffin preparation.
3. To emphasize the roles of sugar, egg, milk, flour, and fat in muffins.

Basic Recipe for Muffins

Ingredients
2 c. flour
½ t. salt
2 ½ t. double-acting baking powder
2 T. sugar
2 T. oil
1 egg
1 c. whole milk
Aerosol vegetable oil cooking spray

Procedure
1. Spray muffin tins with vegetable oil cooking spray or place paper liners in the muffin tins.
2. Sift the dry ingredients together.
3. In a separate bowl, blend egg, milk, and oil.
4. Make a well in the center of the dry ingredients and pour the liquid ingredients into the well, all at once.
5. Stir the batter **only 4 strokes**. Then spoon enough batter for three muffins into the greased muffin pan. Fill each compartment about two-thirds full.
6. Stir the batter **an additional 10 strokes (14 total)**, just enough to dampen all the dry ingredients. The batter will be lumpy. Spoon enough batter for three muffins into the greased muffin pan.
7. Stir the batter **an additional 24 strokes (38 total)**. Spoon enough batter for three muffins into the greased muffin tin.
8. Stir the remaining batter **an additional 32 strokes (70 total)**. Spoon into remaining greased muffin cups. The batter should be smooth and shiny (see Figure 13-4). Be sure to label each stroke variation.
9. Bake at 425 °F for 15 to 20 minutes. Cool on a cooling rack out of the pan for 10 minutes.
10. Slice a muffin down the center, top to bottom, with a sharp utility knife and measure the height. Evaluate the appearance, texture, and flavor of the muffins and document observations in Table D-1.

Characteristics of a high-quality muffin (see Figure 13-5):
- Appearance: top surface is rough and golden brown
- Shape: slightly crowned top, symmetrical in shape
- Interior: has small air cells, all about the same size
- Tenderness: slight resistance to chew

Figure 13-4: Over-stirred muffin batter—long gluten strands, smooth.

Variations

1. All-purpose flour – Follow the basic recipe for muffins as above.

2. Cake flour – Follow the basic recipe for muffins except use cake flour instead of all-purpose flour.

3. Whole-wheat flour – Follow the basic recipe for muffins except use whole-wheat flour instead of all-purpose flour.

Table D-1 Effect of Manipulation and Flour Type on Muffins

Flour type & total strokes	Height	Appearance		Texture	Flavor
		Exterior	Interior		
All-purpose					
4					
14					
38					
70					
Cake					
4					
14					
38					
70					
Whole-wheat					
4					
14					
38					
70					

Figure 13-5: Correctly stirred muffin on the left, and over-stirred on the right; exterior (top) and interior (bottom).

E. Effect of Manipulation on Biscuits

<u>Objectives</u>
1. To emphasize the roles of flour, milk, fat, and baking powder in biscuits.
2. To demonstrate the effect of manipulation on gluten development in biscuit dough.
3. To compare biscuits prepared from a commercial biscuit mix to homemade biscuits.
4. To gain practice in biscuit preparation.

Basic Recipe for Biscuits

Ingredients
1 c. flour
¼ t. salt
1 ½ t. double-acting baking powder
2 T. shortening
6 T. whole milk

Procedure
1. Sift the dry ingredients together.
2. Cut the fat into the dry ingredients with a pastry blender or food processor until the mixture resembles coarse meal.
3. Add milk and stir until ingredients are moistened.
4. Knead the dough ten times on a lightly floured board.
5. Pat or roll the dough until it is ¾" thick.
6. Cut with a floured biscuit cutter and bake the biscuits on an ungreased sheet at 425 °F for 10 to 12 minutes.
7. Measure the tenderness of 3 of the biscuits with a penetrometer. Calculate an average reading and record in Table E-1. Evaluate appearance, flakiness, tenderness and flavor and record observations in Table E-1.

Variations

1. No kneading – Prepare basic recipe for biscuits, but do not knead the dough.
2. Knead 10 times – Prepare the basic recipe for biscuits as above.
3. Kneaded 40 times – Prepare basic recipe for biscuits, but knead the dough 40 times.
4. Prepared with oil – Prepare basic recipe, substituting 2 T. oil for shortening.
5. Prepared with dairy spread – Prepare basic recipe, substituting 2 T. of a soft tube dairy spread for shortening.
6. Commercial biscuit mix – Prepare biscuits from a commercial biscuit mix according to package directions.

Table E-1 Effect of Manipulation on Biscuits

Biscuits	Avg. penetro-meter reading	Appearance	Flakiness	Tenderness (subjective)	Flavor
No kneading					
Kneaded 10 times (control)					
Kneaded 40 times					
Oil					
Soft dairy spread					
Commercial biscuit mix					

Characteristics of a high-quality biscuit:

- Appearance: top is golden brown, sides cream colored
- Texture: layers of flakes are obvious and peel off as pulled apart
- Tenderness: slight resistance to chew
- Flavor: mild to bland depending on type fat used

F. Effect of Sugar on Corn Bread

Objectives

1. To compare corn bread prepared with and without sugar for appearance, texture, and flavor.
2. To emphasize the functions of sugar in quick breads.

Basic Recipe for Corn Bread

Ingredients

1 c. yellow or white corn meal
1 c. all-purpose flour
¼ c. granulated sugar
4 t. double-acting baking powder
½ t. salt
1 c. whole milk
1 egg
¼ c. oil
Aerosol vegetable oil cooking spray

Procedure

1. In a large mixing bowl, combine the corn meal, flour, sugar, baking powder, and salt. Mix well.
2. In a separate bowl, combine the milk, egg, and oil. Mix well.
3. Add the wet ingredients to the combined dry ingredients and beat until fairly smooth.
4. Spray 8" × 8" pan with vegetable oil cooking spray. Place cornbread mixture in prepared baking pan and bake in a preheated 425 °F oven for 20 to 25 minutes.
5. Evaluate the appearance, texture, and flavor of the corn bread and record observations in Table F-1.

Variations

1. Sugar – Prepare basic recipe for cornbread.

2. No sugar – Prepare basic recipe for cornbread, except omit the sugar.

3. Baking soda + buttermilk – Prepare the basic recipe for cornbread, substituting ½ t. baking soda for the baking powder and 1 c. buttermilk for the whole milk.

4. Baking soda + buttermilk, no sugar – Prepare variation #3, except omit the sugar.

Table F-1 Effect of Sugar on Cornbread

Type of Cornbread	Appearance	Texture	Flavor
1. Sugar			
2. No sugar			
3. Soda + buttermilk			
4. Soda + buttermilk, no sugar			

Additional Food Preparation Exercises

G. Leavening by Steam: Cream Puff Shells

Objectives

1. To demonstrate the use of steam as a primary leavening agent in cream puffs.
2. To emphasize the importance of baking temperature for steam-leavened products.

Cream Puff Shells

Ingredients

1 c. water
½ c. butter
¼ t. salt
1 c. flour
4 eggs
Aerosol vegetable oil cooking spray

Procedure

1. In a saucepan, bring the water, butter, and salt to a boil. Be sure that the butter is completely melted.
2. Add flour to the saucepan all at once and stir well.
3. Cook over medium heat until the mixture forms a cohesive ball that leaves the side of the pan. Stir continuously with a wooden spoon.
4. After the mixture has cooled slightly, add the eggs, one at a time.
5. After each egg is added, beat the mixture very well. It should gradually become smooth.
6. Spray the baking sheet with vegetable oil cooking spray. Drop the cream puff paste onto the prepared baking sheet.
7. Bake in a 450 °F oven for 15-20 minutes or until cream puffs appear to reach their maximum volume. Reduce oven temperature to 325 °F and bake for another 25-30 minutes or until cream puffs are well browned and interiors are dry.

H. Preparation of Various Quick Breads

Objectives

1. To compare the ingredients, methods of mixing, and characteristics of various types of quick breads.
2. To observe the changes that occur during the mixing and baking of quick breads.
3. To gain experience in the preparation of quick breads.

H-a. Banana Bread

Banana Bread

Ingredients

1 c. all-purpose flour
¾ c. whole-wheat flour
¼ c. wheat germ
¼ c. roasted sunflower seeds (optional)
1 T. double-acting baking powder
$^1/_3$ c. milk
1 c. mashed ripe bananas
2 eggs, slightly beaten
½ c. honey
4 T. oil
Aerosol vegetable oil cooking spray

Procedure

1. Combine flours, seeds, wheat germ, and baking powder in a large mixing bowl.
2. In a separate bowl, mix together bananas, eggs, honey, oil, and milk.
3. Add the wet mixture to the dry mixture and stir only until ingredients are just blended.
4. Spray 9" × 5" loaf pan with vegetable oil cooking spray. Pour batter into prepared pan and bake at 350 °F for 55 to 60 minutes or until toothpick inserted in center comes out clean.

H-b. Raisin-Bran Muffins

Raisin-Bran Muffins

Ingredients

1 c. all-purpose flour
1 c. bran cereal
2 t. double-acting baking powder
1 egg, slightly beaten
2 T. oil
2 T. honey
¼ c. molasses
$^1/_3$ c. raisins
1 c. buttermilk
Aerosol vegetable oil cooking spray

Procedure

1. Combine bran and milk. Let stand 5 minutes.
2. Sift flour with baking powder. Mix in the raisins.
3. In a separate bowl, mix together egg, oil, honey, molasses, and the bran-milk mixture.
4. Add the wet mixture to the dry mixture and stir only until ingredients are just blended.
5. Spray muffin tin with vegetable oil cooking spray. Spoon batter into prepared muffin tins, filling $^2/_3$ full, and bake at 400° F for 20 to 25 minutes. Makes 6 muffins.

H-c. Cranberry Pumpkin Bread

Cranberry Pumpkin Bread

Ingredients

2 eggs
¼ c. vegetable oil
1 ½ c. sugar
8 oz. canned pumpkin
1 ¾ c. 2 T. flour
1 t. baking soda
½ t. salt
½ t. cinnamon
¼ t. ground nutmeg
⅛ t. ground cloves
⅛ t. ground ginger
1 c. whole fresh cranberries
½ c. pecans (optional)
Aerosol vegetable oil cooking spray

Procedure

1. Beat eggs in a large mixing bowl. Blend in oil, sugar, and pumpkin.
2. Sift dry ingredients including spices together in a separate bowl.
3. Add dry ingredients to pumpkin mixture.
4. Stir in cranberries and pecans (if desired).
5. Spray 9" × 5" loaf pan with vegetable oil cooking spray. Pour batter into prepared pan and bake at 350 °F for 60 minutes or until toothpick inserted in center comes out clean.
6. Let cool approximately 5 minutes before transferring bread from baking pan to cooling rack.

From the recipe file of Karen Beathard

H-d. Date Nut Bread

Date Nut Bread

Ingredients

8 oz. dates
1 t. baking soda
1 c. boiling water
1 egg, beaten
1 c. sugar
2 c. 2 T. flour, divided
1 t. baking powder
½ t. salt
1 t. vanilla extract
½ c. chopped pecans
Shortening

Procedure

1. Preheat oven at 325 °F.
2. Mix dates, baking soda, and boiling water in a large mixing bowl and allow to sit for 2-3 minutes.
3. Meanwhile, sift 2 c. flour, baking powder, and salt together and set aside.
4. Add beaten egg and sugar to date mixture.
5. Mix in sifted dry ingredients and vanilla. Stir in pecans.
6. Use shortening and 2 T. flour to grease and flour a 9" × 5" loaf pan. Pour batter into prepared pan and bake at 325 °F for 55 to 60 minutes or until toothpick inserted in center comes out clean.

From the recipe file of Alma Mynhier

H-e. Pancakes

Pancakes

Ingredients

1 ¼ c. flour
½ t. salt
1 T. baking powder
1 T. sugar
1 egg
1 c. milk
2 T. oil

Procedure

1. Sift dry ingredients together to ensure complete mixing.
2. In a separate bowl, combine milk, eggs, and oil and mix well.
3. Add liquid ingredients to dry ingredients all at once; stir to combine. Batter may appear lumpy. Do not beat or mix until smooth. Batter should be fairly heavy and not runny.
4. If batter seems too thick, add a small amount of milk and mix well.
5. Cook pancakes on a hot, lightly oiled griddle or fry pan.
6. Cook 1-2 minutes on the second side.
7. Keep warm in oven until served.

From the recipe file of Lesha Emerson

Post-Lab Study Questions

Textbook reference: Chapters 12, 17, and 19

Discussion Questions

1. What chemical leaveners were used in the muffins? Why are they used in quick breads?

2. Why are steam-leavened products often baked at two temperatures?

3. What is the secondary leavening agent in popovers and cream puffs?

4. What functions do the eggs have in these recipes?

5. Why do cream puffs require 4 eggs but popovers only 2?

Questions for Post-Lab Writing

6. What conditions are required for baking soda and baking powder to react and give off carbon dioxide?

7. In the preparation of angel food cake, why was part of the sugar incorporated into the egg foam?

8. Why was cream of tartar added to the egg foam?

9. Why was the flour-sugar mixture folded into the egg white foam, not beaten in?

10. How did beating the angel cake batter with an electric mixer affect the baked product volume? The product appearance?

11. How did the baking temperature of the angel cake affect the product volume? The product appearance?

12. What is the primary leavening agent in popovers? How does this influence the baking temperature required?

13. Compare the popovers prepared with two eggs to the popovers prepared with one egg. Is there a difference? If so, what do you think caused this difference?

14. Describe a high-quality muffin. How does the amount of mixing influence muffin quality?

15. Compare the muffins prepared with different types of flour. Which flour type was most affected by over-mixing? Why?

16. Why are biscuits kneaded?

17. What effect does over-kneading have on tenderness and volume of biscuits?

18. Does the size of the fat particles in biscuits have any effect on the finished product? Does oil give the same result as solid fat?

19. Compare the biscuits made with commercial mix with the biscuits prepared from “scratch.”

20. Would you expect biscuits prepared by different cooks using the same recipe to be different? Why?

21. Compare the corn bread prepared with and without sugar. What are the functions of sugar in this recipe?

22. Is excess gluten formation a problem in corn bread? Why or why not?

Unit 14 – Yeast Breads

Introduction

The essential ingredients for baking bread are yeast, flour, liquid, and salt. The yeast that is used in bread-making is *Saccharomyces cerevisiae*, commonly called baker's yeast. The primary function of the yeast is to leaven the dough by producing carbon dioxide. Carbon dioxide is formed as a result of the fermentation of sugars naturally present in the flour. Additional sugar may be added to increase the fermentation rate. Rapid-rise yeast can also be used; generally, rising times are cut in half. Flour also provides the proteins from which gluten is developed, glutenins and gliadins. Liquid in bread is necessary to hydrate proteins and starch and is essential for gluten development. Salt has a stabilizing influence on yeast fermentation. Optional ingredients include fat, sugar, eggs, and additives. The fermentation temperature used for the yeast is important in the successful preparation of breads. The optimum temperature is 80 °F. Too high a temperature causes rapid production of carbon dioxide and the dough rises too quickly. A sour off-flavor results due to the formation of undesirable by-products of fermentation. Too low a temperature causes inadequate production of carbon dioxide and the dough is slow to rise. Bread dough usually is kneaded. This is necessary to mix the ingredients, to form numerous gas cells, and to develop gluten, which gives the dough elasticity and gas-holding capacity.

Bread dough may be prepared by the straight dough method, the sponge method, or the batter method. In the straight dough method, any fat, yeast, sugar, and salt used are dispersed in the liquid. All of the flour is gradually added to the liquid mixture until soft dough is formed. The dough is kneaded to develop gluten, which is necessary for the retention of gas during fermentation and proofing. The sponge method of mixing dough involves combining all of the liquid with the yeast and part of the sugar and flour. The mixture is allowed to ferment for a half-hour to an hour to become foamy and spongy before the rest of the ingredients are added and the dough is kneaded. The batter method is the simplest mixing technique. All ingredients are combined and beaten by hand or electric mixer to develop the gluten. The batter is ready when it no longer sticks to the side of the bowl. Breads prepared by this method often have a yeasty flavor and a coarse grain.

Key Terms

dissolve – to mix a solid dry substance with a liquid until the solid substance is in solution.

fermentation – the increase in size that the dough experiences as a result of carbon dioxide production and enzyme and pH alterations.

oven spring – the quick expansion of dough during the first ten minutes of baking, which is due to expanding gases.

proof – the rising or increase in volume of shaped dough that occurs through fermentation.

proof box – equipment designed to maintain optimal temperatures and humidity for the fermentation and rising of a yeast dough.

Pre-Lab Questions

Textbook reference: Chapter 20

1. What are the essential ingredients in yeast bread? Identify the roles of these ingredients.

2. Why is the dough fermented and proofed? At what temperature should fermentation take place, ideally?

3. What occurs during the fermentation of yeast bread?

4. What are the objectives in punching down the dough?

5. Why is yeast dough kneaded?

6. Explain the procedures of the straight dough mixing method.

Lab Procedures

A. Preparation of Yeast Breads by Various Methods

Objectives

1. To gain practice in the procedure for making yeast dough by the straight dough and batter methods.
2. To emphasize the importance of kneading and observe the effects of over-kneading and under-kneading yeast breads.
3. To observe the effect of salt on yeast bread.
4. To observe the changes that occur during the various stages of bread dough preparation.
5. To gain practice in the preparation of yeast bread.

Basic Recipe for Yeast Bread

Ingredients

½ t. salt
1 envelope active dry yeast
2 T. water
1 ¾ c. bread flour (approximately)
½ c. whole milk
2 t. granulated sugar
2 t. shortening
Aerosol vegetable oil cooking spray

Procedure

1. Dissolve ½ t. sugar in 2 T. lukewarm water (104 °F or 40 °C). Sprinkle one envelope active dry yeast over this mixture. **Do not stir.** Allow the yeast to soften for 10 minutes.
2. In a saucepan, heat the milk on medium to scalding (140 °F or 60 °C). Add the remaining sugar, shortening, and salt.
3. Cool milk to room temperature and transfer to a large mixing bowl. Add the yeast mixture and stir well.
4. Stir in half of the flour. Mix well.
5. Keep adding flour until the dough loses enough of its stickiness to allow kneading. If you have added all of the flour and the dough still seems sticky, check with the instructor.
6. Lightly flour a clean, sanitized board or countertop. Knead the dough for 15 minutes or until the dough feels smooth and elastic.
7. Spray the mixing bowl with aerosol vegetable oil cooking spray. Return dough to greased mixing bowl, turning to grease top, and cover with waxed paper or a towel. Place the dough in a proof box or a warm place (80 °F) until doubled in size.
8. Gently punch down dough. Gather the dough into a ball, manipulating slightly, if necessary, to be sure the dough is homogeneous. Shape the dough into a loaf. Place the loaf in a greased 9 ¼" × 5 ¼" × 2 ¾" loaf pan. Cover the loaf with wax paper or a clean towel to keep the surface from drying. The surface of the loaf may be lightly greased, if desired.
9. Place the loaf in a proof box or a warm place (80 °F) until doubled in size.
10. Bake the loaf at 400 °F for about 30 minutes.
11. Evaluate the appearance, texture, and flavor of the bread, and record observations in Table A-1.

Characteristics of a high-quality yeast bread:

- Exterior: surface is light- to medium-brown, smooth
- Interior: air cells are small and evenly sized throughout the loaf; cell walls are delicate and thin
- Tenderness: there is little resistance when bitten into; only requires 4 to 5 chews for swallowing
- Flavor: bland with only a hint of yeast flavor

Variations

1. Control – Follow the basic recipe for yeast bread as above.
2. Knead 3 minutes – Follow the basic recipe for yeast bread, but knead the dough for only 3 minutes instead of 15 minutes.
3. Knead 30 minutes – Follow the basic recipe for yeast bread, but knead the dough for 30 minutes instead of 15 minutes.
4. No salt – Follow the basic recipe for yeast bread, but omit the salt.
5. Extra salt – Follow the basic recipe for yeast bread, but use 3 t. salt.

6. Bread maker – Prepare yeast bread according to the manufacturer recipe for bread in a bread maker.
7. Batter method – Follow the basic recipe for yeast rolls using the batter method below.

Basic Recipe for Yeast Rolls Using the Batter Method

Ingredients
¾ c. milk
3 T. granulated sugar
1 t. salt
¼ c. shortening
1 ½ c. lukewarm water (~104 °F or ~40 °C)
2 t. granulated sugar
2 envelopes active dry yeast
2 c. sifted bread flour
Aerosol vegetable oil cooking spray

Procedure
1. Heat milk to scalding (140 °F) in a saucepan over medium heat.
2. Stir in 3 T. sugar, salt, and shortening. Cool to lukewarm.
3. Meanwhile, dissolve 2 t. sugar in ½ c. lukewarm water (104 °F or 40 °C) in a large mixing bowl. Sprinkle 2 envelopes active dry yeast over this mixture and let stand 10 minutes. Then stir well.
4. Stir the lukewarm milk mixture into the yeast mixture. Add 2 c. sifted bread flour. Stir until well blended.
5. Cover with a towel and let rise in a proof box or a warm place (80 °F or 27 °C) until doubled in bulk (about 40 minutes).
6. Stir down the batter. Fill greased muffin pans about ¾ full.
7. Bake in a hot oven at 400 °F for 20-25 minutes.
8. Evaluate the appearance, texture, and flavor of the rolls, and record this in Table A-1.

Table A-1 Preparation of Yeast Breads by Various Methods

Yeast Bread/Rolls	Appearance	Texture	Flavor
Basic recipe (control)			
Knead 3 minutes			
Knead 30 minutes			
No salt			
Extra salt bread			
Bread from bread maker			
Batter method: yeast rolls			

Additional Food Preparation Exercises

B. Preparation of Various Yeast Breads

Objectives

1. To prepare various types of yeast bread products and compare their ingredients and methods of mixing.
2. To demonstrate the proper techniques for preparing yeast breads.

B-a. Easy Yeast Rolls

Easy Yeast Rolls

Ingredients

3 ½ to 4 ½ c. flour, unsifted (bread or all-purpose)
3 T. sugar
1 t. salt
2 packages active dry yeast
1 c. whole milk
½ c. water
¼ c. butter or margarine
Aerosol vegetable oil cooking spray

Procedure

1. Using a wooden spoon, mix together 1 ½ c. flour, sugar, salt, and yeast in a large mixing bowl.
2. Combine the milk, water, and margarine in a saucepan and heat over low heat until warm (120 °F to 130 °F or 49 °C to 54 °C). Do not overheat. The margarine does not need to melt.
3. Gradually add the liquid ingredients to the dry ingredients. Beat two minutes at the medium speed of an electric mixer. Scrape bowl occasionally.
4. Add ½ c. flour and beat at high speed for two minutes. Scrape bowl occasionally.
5. Add enough additional flour to make a soft dough.
6. Turn out onto a lightly floured, clean, sanitized board or countertop. Knead until smooth and elastic, about 5 minutes.
7. Turn on oven to 250 °F for one minute, and then **turn off**.
8. Place dough into a greased bowl. Turn the dough over to grease the top.
9. Cover the dough with waxed paper or a clean towel. Place it in the warm oven and let it rise for 15 minutes.
10. Punch down the dough. Shape it into various kinds of rolls, braids, and other shapes as directed by the instructor and place on baking sheet. (Note: this dough does not do well as full-size loaves of bread.)
11. Turn on the oven again to 250 °F for one minute, and then **turn off**.
12. Cover the dough, place it in the warm oven, and then let it proof for 15 minutes.
13. Remove the dough from the oven and heat the oven to 425 °F. Bake the rolls or other shaped dough, uncovered, for 10-12 minutes or until lightly browned.

B-b. Whole-Wheat Bread

Whole-Wheat Bread

Ingredients

½ t. salt
1 envelope active dry yeast
2 T. water
$^{7}/_{8}$ c. bread flour (approximately)
$^{7}/_{8}$ c. whole-wheat flour (approximately)
½ c. milk
2 t. granulated sugar
2 t. shortening
Aerosol vegetable oil cooking spray

Procedure

1. Dissolve ½ t. sugar in 2 T. of lukewarm water (104 °F). Sprinkle 1 envelope active dry yeast over this mixture. **Do not stir.** Allow the yeast to soften for 10 minutes.
2. Heat the milk to scalding (198 °F) in a saucepan and add the remaining sugar, shortening, and salt.
3. Cool milk to room temperature and transfer to a mixing bowl. Add the yeast mixture and stir well.
4. Stir in half of the flour. Mix well.
5. Keep adding flour until the dough loses enough of its stickiness to allow kneading. If you have added all of the flour and the dough still seems sticky, check with the instructor.
6. **Lightly** flour a clean, sanitized board or countertop. Knead the dough for 15 minutes or until the dough feels smooth and elastic.
7. Spray the mixing bowl with aerosol vegetable oil cooking spray. Return dough to greased mixing bowl, turning to grease top, and cover with wax paper or a towel. Place the dough in a proof box or a warm place (80 °F) until doubled in size.
8. **Gently** punch down dough. Gather the dough into a ball, manipulating slightly, if necessary, to be sure the dough is homogeneous. Shape the dough into a loaf.
9. Place the loaf in a lightly greased loaf pan (8 ½" × 4 ½" × 3"). Cover the loaf with waxed paper or a clean towel to keep the surface from drying. The surface of the loaf may be lightly greased, if desired.
10. Place the loaf in a proof box or a warm place (80 °F) until doubled in size.
11. Bake the loaf at 400 °F for about 30 minutes.

B-c. Rye Bread

Rye Bread

Ingredients

½ t. salt
1 envelope active dry yeast
2 T. water
$^{7}/_{8}$ c. bread flour (approximately)
$^{7}/_{8}$ c. rye flour (approximately)
1-2 T. caraway seeds
½ c. milk
2 t. granulated sugar
2 t. shortening
Aerosol vegetable oil cooking spray

Procedure

1. Dissolve ½ t. sugar in 2 T. lukewarm water (104 °F). Sprinkle one envelope active dry yeast over this mixture. **Do not stir.** Allow the yeast to soften for 10 minutes.
2. In a saucepan, heat the milk to scalding (140 °F) over medium heat. Add the remaining sugar, shortening, and salt.
3. Cool milk to room temperature and transfer to a large mixing bowl. Add the yeast mixture and stir well.
4. Stir in half of the flour. Mix well.
5. Keep adding flour until the dough loses enough of its stickiness to allow kneading. If you have added all of the flour and the dough still seems sticky, check with the instructor.
6. **Lightly** flour a clean, sanitized board or countertop. Knead the dough for 15 minutes or until the dough feels smooth and elastic.
7. Spray the mixing bowl with aerosol vegetable oil cooking spray. Return dough to greased mixing bowl, turning to grease top, and cover with waxed paper or a towel. Place the dough in a proof box or a warm place (80 °F) until doubled in size.
8. **Gently** punch down dough. Gather the dough into a ball, manipulating slightly, if necessary, to be sure the dough is homogeneous. Shape the dough into a loaf. Place the loaf in a greased loaf pan. Cover the loaf with waxed paper or a clean towel to keep the surface from drying. The surface of the loaf may be lightly greased, if desired.
9. Place the loaf in a proof box or a warm place (80 °F) until doubled in size.
10. Bake the loaf at 400 °F for about 30 minutes.

B-d. Cinnamon Rolls

Cinnamon Rolls

Ingredients

½ t. salt
1 envelope active dry yeast
2 T. water
1 ¾ c. bread flour (approximately)
½ c. 2 T. milk, divided
4 T. granulated sugar, divided
2 T. shortening
1 t. melted butter
1 t. cinnamon
2 T. nuts
2 T. raisins
1 c. confectioners' sugar, sifted
Aerosol vegetable oil cooking spray

Procedure

1. Dissolve ½ t. sugar in 2 T. lukewarm water (104 °F). Sprinkle one envelope active dry yeast over this mixture. **Do not stir.** Allow the yeast to soften for 10 minutes.
2. In a saucepan, heat ½ c. milk to scalding (140 °F). Add the remaining sugar, shortening, and salt.
3. Cool milk to room temperature and transfer to a large mixing bowl. Add the yeast mixture and stir well.
4. Stir in half of the flour. Mix well.
5. Keep adding flour until the dough loses enough of its stickiness to allow kneading. If you have added all of the flour and the dough still seems sticky, check with the instructor.
6. **Lightly** flour a clean, sanitized board or countertop. Knead the dough for 15 minutes or until the dough feels smooth and elastic.
7. Spray the mixing bowl with aerosol vegetable oil cooking spray. Return dough to greased mixing bowl, turning to grease top, and cover with waxed paper or a towel. Place the dough in a proof box or a warm place (80 °F) until doubled in size.
8. Roll dough into a rectangle.
9. Brush with 1 t. melted butter. Sprinkle with 1 t. cinnamon and 2 T. each of nuts, raisins, and sugar.
10. Roll up as for a jelly roll and slice the roll into ½" thick slices.
11. Place in greased pan and let rise until approximately doubled in volume.
12. Bake at 375 °F for 20-30 minutes.
13. To prepare glaze, sift 1 c. confectioners' sugar. Use a whisk to mix sifted confectioners' sugar with 2 T. milk.
14. Top with glaze while the rolls are still warm.

B-e. Pita Bread

Pita Bread

Ingredients

½ package active dry yeast
$^{2}/_{3}$ c. lukewarm water
½ T. honey
½ t. salt
2 c. all-purpose flour
1 T. oil

Procedure

1. Mix ¼ c. lukewarm water (104 °F) with ½ t. honey in a mixing bowl. Sprinkle ½ package active dry yeast over this mixture and let stand 5 minutes. Then stir.
2. Add the remaining honey, flour, oil, and salt and mix with a wooden spoon until well-combined.
3. Turn out onto a lightly floured, clean, sanitized board or countertop and knead for 10-15 minutes until the dough is smooth and elastic.
4. Oil the mixing bowl, return dough to bowl, and lightly oil the top surface of the dough. Cover and let rise 45-60 minutes or until the dough is smooth and elastic.
5. Punch the dough down and knead again for a few minutes. Divide dough into 4 equal parts. Form each part into a smooth round ball. Cover the balls with a clean towel and let rest 15 minutes.
6. Preheat oven to 500 °F.
7. Roll each ball into round loaves $^{1}/_{8}$" thick. Place on a lightly floured baking sheet. Cover with a towel and let the loaves rest for 30 minutes. Bake breads on the lowest oven rack for 10 minutes, or until they are puffed up and brown.
8. Wrap the freshly-baked breads in foil for 15 minutes. Then unwrap. The loaves will deflate and will have a pocket in the center.

Post-Lab Study Questions

Textbook reference: Chapters 17 and 20

Discussion Questions

1. Why is active dry yeast sprinkled on water and allowed to sit before stirring? Why is sugar added to the water?

2. What is the purpose of scalding the milk? What would be the result of adding the yeast to the scalded milk before the milk was cooled?

3. Compare the texture and flavor of the no-knead rolls with bread prepared by the straight-dough method. Explain any differences.

Questions for Post-Lab Writing

4. What will the finished product be like if the dough is under-kneaded? Over-kneaded? Why?

5. How does salt affect yeast dough? What will the finished product be like if salt is omitted? If too much salt is used?

6. Why were the rising times so short for the easy yeast rolls?

7. What are some possible sources of error in making yeast bread?

Unit 15 – Fats and Oils

Introduction

A wide variety of fats and oils are available in the marketplace. They are used in numerous food products, such as spreads for breads and bread products, shortening or tenderizing ingredients in baked products, ingredients in emulsions, flavoring agents, a source of energy, and a cooking medium in deep-fat frying.

Fats do not boil. At high temperatures, fats decompose and may form a blue smoke. When this occurs, fats have reached their "smoke point." Smoke points of fats or oils used for high-heat cooking, such as deep-fat frying, must be high enough to prevent decomposition of the fats during use. The smoke point is lowered by the presence of low-molecular weight fatty acids in triglycerides and by the presence of emulsifiers, free fatty acids, and food particles in the fat.

Figure 15-1: Equipment setup used to measure a fat's smoke point.

When deep-fat frying foods, the amount of fat that is absorbed by the food product should be minimized. Fat absorption is increased by increasing the contact time of the food and the fat, lack of a coating or breading on the food to form a physical barrier against the fat, increasing the surface area of food exposed to the fat, and increasing the proportion of fat, sugar, or egg yolk in a batter. Increased fat absorption results in a less desirable fried product. Cooking methods such as sautéing and stir frying generally minimize the amount of fat absorbed by the foods.

In many baked products, substitutions can be made for part or all of the fats and oils. Common substitutes used by consumers include fruit purees such as applesauce. The use of flavored oils, such as olive and sesame oils, is desirable for some products.

Emulsions

An emulsion is formed when one liquid is dispersed in another liquid with which it is ordinarily immiscible. Emulsions may be temporary, semi-permanent, or permanent, depending on their stability. The stability of an emulsion is dependent on the particle size of the dispersed liquid and the type of emulsifier present. An effective emulsifier is a molecule that has both polar and non-polar ends. The non-polar end dissolves in the lipid phase, and the polar end dissolves in the water phase of the emulsion. Emulsifiers aid in forming an emulsion by decreasing the surface tension of the liquids and preventing the coalescence of dispersed fat droplets.

Emulsion instability is displayed by the separation of the two phases, i.e., breaking of the emulsion. Emulsion stability may be sensitive to temperature fluctuations. A mayonnaise emulsion may break due to over-beating, adding too much oil at one time during its formation, surface drying, freezing, or heating. A broken mayonnaise emulsion may be re-emulsified by adding the broken emulsion slowly to an egg yolk while beating.

Egg yolk and milk are naturally occurring emulsions. Emulsions are also formed during the processing and preparation of many food products. These include peanut butter, margarine, whipped topping, icings, puddings, sauces, gravies, mayonnaise, salad dressings, frozen desserts, and beverages. Emulsions also are formed in sausages and baked products.

Key Terms

clarified butter – butter that is treated to remove milk solids and water in order to reduce the potential for burning and increase the smoke point.

hydrogenation – commercial process used to make fats and oils more solid and extend their shelf life. In this process, hydrogen atoms are added to the double bonds in monounsaturated and polyunsaturated fatty acids to make them more saturated.

julienne – to slice food, usually vegetables, into 1" to 3" sticks that are $^{1}/_{16}$" to $^{1}/_{8}$" thick.

rancid – describes spoiled (oxidized) fats or oils; rancidity results in off tastes and odors.

smoke point – the temperature at which a fat or an oil begins to smoke.

stabilizer – ingredient added to food to minimize deterioration and loss of desirable properties.

Pre-Lab Questions

Textbook reference: Chapters 5, and 22

1. Explain the differences among butter, margarine, shortening, and vegetable oil. Compare their nutrient contributions.

2. What is the purpose of an emulsifier?

3. Explain the differences among a temporary, semi-permanent, and permanent emulsion. Give an example of each.

4. What fats are best to use when deep frying? Why?

5. How is energy transferred to foods fried in deep-fat?

Lab Procedures

A. Comparison of Selected Commercial Fat Products Used as Spreads

Objectives

1. To emphasize differences in composition of selected spreads.
2. To compare the spreadability of selected spreads.
3. To compare the flavor of selected spreads.

Basic Procedure to Compare Commercial Fat Products

Ingredients

Assortment of spreads (as chosen by instructor)
Unsalted crackers

Procedure

1. Obtain samples of each commercial spread available.
2. Chill half of each spread for one-half hour in the refrigerator. Keep the rest of each spread at room temperature.
3. Spread ½ t. of each chilled spread and each room-temperature spread on unsalted crackers. Note the "spreadability" of each product.
4. Taste each product, chilled and at room temperature. Compare flavor and mouthfeel and record observations in Table A-1.
5. Using the labeling on the packages as a guide, compare the ingredients and nutritional value. Record this information in Table A-1.

Table A-1 Comparison of Selected Commercial Fat Products Used as Spreads					
Product	Fat composition	Spreadability		Sensory evaluation	
		Chilled	Room temp	Chilled	Room temp

B. Effect of Coating on Fat Absorption

Objectives

1. To demonstrate the effect of a coating on fat absorption by squash during deep-fat frying.
2. To gain experience in using correct deep-fat frying techniques.

Basic Recipe for Fried Yellow Squash

Ingredients

1 medium-sized yellow squash
½ c. flour
½ t. salt
¼ t. pepper
1 egg
2 c. soybean oil

Procedure

1. Wash the squash. Do not peel.
2. Slice the squash into ½" thick circles.
3. Combine the flour, salt, and pepper in a mixing bowl.
4. Beat the egg in a separate bowl and set aside.
5. Heat the oil in a deep-fat fryer (or a 1-qt. saucepan) to 375 °F.
6. Dredge the squash circles into the flour mixture, then into the beaten egg and then again into the flour mixture. Shake off excess flour.
7. Fry the coated squash until lightly browned. Pieces should float freely. Do not overload the pan. Note the frying time.
8. Drain fried squash on paper towels.
9. Note appearance, texture, flavor, and mouthfeel and record observations in Table B-1.

Characteristics of a high-quality fried vegetable:

- Appearance: golden brown, not oily
- Texture: crisp outer surface, soft center
- Flavor: mild flavor of the vegetable, not oily

Variations

1. Breaded – Follow the basic recipe for fried squash as above.

2. Unbreaded – Follow the basic recipe for fried squash, but do not bread the squash. Fry unbreaded squash at 375 °F for the same amount of time that the coated squash circles are fried.

Table B-1 Effect of Coating on Fat Absorption				
Type of squash	**Appearance**	**Texture**	**Flavor**	**Mouthfeel**
Breaded				
Unbreaded				

C. *Effect of Cooking Time on Fat Absorption*

Objectives

1. To demonstrate the influence of cooking time on fat absorption by donut holes.
2. To gain experience in using correct deep-fat frying techniques.

Basic Recipe for Frying Canned Biscuit "Donut Holes"

Ingredients

1 refrigerated canned biscuit (uncooked)
2 c. soybean oil

Procedure

1. Cut a biscuit into fourths. Weigh **this group of biscuit dough pieces** and record the group weight in Table C-1.
2. Heat the oil in a deep-fat fryer (or 1-qt. saucepan) to 375 °F (190 °C).
3. Fry biscuits for 2 minutes (1 minute per side).
4. Drain biscuits on paper towels after removing from the hot oil. Label the group based on cooking time. Allow them to cool to ~90 °F (~32 °C) and then reweigh **the group** of biscuit donut holes.
5. Calculate both the weight gained and the percentage of weight gained for each group and record in Table C-1.

Variations

1. 2 minutes – Follow basic recipe for donut holes as above.

2. 4 minutes – Follow basic recipe for donut holes, except fry the biscuit pieces for 4 minutes (2 minutes per side).

3. 6 minutes – Follow basic recipe for donut holes, except fry the biscuit pieces for 6 minutes (3 minutes per side).

4. 10 minutes – Follow basic recipe for donut holes, except fry the biscuit pieces for 10 minutes (5 minutes per side).

5. 2, 4, 6, or 10 minutes at 350 °F – Follow the instructions for variations 1-4, except heat the frying oil to 350 °F (177 °C) and fry the biscuit pieces at that temperature. Record observations in Table C-2.

Table C-1 Effect of Cooking Time on Fat Absorption at 375 °F

Cooking time (minutes)	Weight before cooking	Weight after cooking	Weight gain	% weight gain
2				
4				
6				
10				

Table C-2 Effect of Cooking Time on Fat Absorption at 350 °F

Cooking time (minutes)	Weight before cooking	Weight after cooking	Weight gain	% weight gain
2				
4				
6				
10				

D. Comparing Stir-Fry and Oven Fry Cooking Methods

Objectives

1. To compare methods of cooking with fats and oils.
2. To gain practice in stir-frying and oven frying.

Basic Recipe for Stir-Fried Zucchini

Ingredients

2 T. oil
2 medium fresh zucchini
1 small onion, chopped
¼ t. garlic powder
¼ t. salt

Procedure

1. Wash zucchini with a vegetable brush. Remove ends of zucchini.
2. Julienne zucchini into 3" long × ¼" wide strips.
3. Heat oil in a skillet on medium high. Test oil temperature by dropping 1 piece of zucchini into the oil. (If it sizzles, the oil is hot enough.)
4. Add zucchini, onion, garlic powder, and salt and cook with occasional stirring.
5. Stir fry for about 4 minutes, only until tender.
6. Transfer onto display plate.
7. Observe the appearance, flavor, and texture and record observations in Table D-1.

Variations

1. Stir-fry – Prepare zucchini according to the basic recipe.

2. Oven fry – Preheat oven to 400 °F. Wash and julienne zucchini according to basic recipe. Mix ½ c. parmesan cheese, ½ c. fine bread crumbs, and ¼ t. salt and set aside. Beat 1 large egg in another bowl. Prepare baking pan by spraying with aerosol vegetable oil cooking spray. Dip zucchini in beaten egg to coat. Then dip the zucchini in the parmesan cheese and bread crumb mixture. Place the breaded zucchini on the prepared baking pan. Bake for 8 minutes at 400 °F. Increase temperature to broil and cook for an additional 1-2 minutes until golden brown. Remove zucchini from oven and transfer to display plate for comparison.

Table D-1 Comparing Stir-Fry and Oven Fry Cooking Methods

Method of cooking	Appearance	Flavor	Texture
Stir fry			
Oven fry			

E. Polyunsaturated versus Monounsaturated Oils: Vegetable Oil and Extra Virgin Olive Oil

Objectives
1. To evaluate differences between polyunsaturated and monounsaturated oils in food preparation.
2. To gain experience in making salad dressings.

Basic Recipe for Cucumber-Tomato Salad

Ingredients for Salad
½ lb. cucumbers
½ lb. tomatoes
½ red onion

Procedure
1. Wash and slice cucumbers to ¼" thickness.
2. Slice tomatoes and onions on cutting board.
3. Place in large bowl.

Basic Recipe for Oil and Vinegar Dressing

Ingredients for Dressing
Pinch black pepper
½ t. salt
½ c. vegetable oil
2 T. vinegar

Procedure
1. Combine salt and pepper in a mixing bowl.
2. Add vegetable oil and vinegar alternately, mixing until thoroughly combined.
3. Pour over salad.
4. Evaluate appearance, flavor, and texture and record observations in Table E-1.

Variations

1. Vegetable oil – Follow basic recipe for salad and oil and vinegar dressing as above.

2. Extra virgin olive oil – Follow basic recipe for salad and oil and vinegar dressing, except substitute ½ c. extra virgin olive oil for vegetable oil.

Table E-1 Polyunsaturated versus Monounsaturated Oils: Vegetable Oil and Olive Oil			
Salad dressing	**Appearance**	**Flavor**	**Texture**
Vegetable oil			
Olive oil			

F. Comparing Fat and Fat Substitutes

Objectives
1. To evaluate differences between foods prepared with fat and with a fat substitute.
2. To gain practice in preparing baked products.

Basic Recipe for Brownies

Ingredients

$^{1}/_{3}$ c. melted margarine
¼ c. cocoa
1 c. sugar
2 eggs
$^{2}/_{3}$ c. flour
$^{2}/_{3}$ t. vanilla
Aerosol vegetable oil cooking spray

Procedure
1. Preheat oven to 350° F.
2. Lightly grease an 8" pan with aerosol vegetable oil cooking spray.
3. Measure cocoa into a mixing bowl.
4. Melt margarine. Add to cocoa and mix well.
5. Add eggs, sugar, flour, and vanilla.
6. Pour into 8" prepared pan. Bake for approximately 25-30 minutes, checking periodically for doneness.
7. Evaluate the brownies for appearance, flavor, and texture and record observations in Table F-1.

Variations

1. Margarine – Follow basic recipe for brownies as above.

2. Applesauce – Follow basic recipe for brownies, except substitute $^{1}/_{3}$ c. applesauce for melted margarine.

3. Baby food prunes – Follow basic recipe for brownies, except substitute $^{1}/_{3}$ c. baby food prunes for melted margarine.

Table F-1 Comparing Fat and Fat Substitutes			
Fat or fat substitute	**Appearance**	**Flavor**	**Texture**
Margarine			
Applesauce			
Baby food prunes			

G. Effect of Solid versus Liquid Fat on Quality of a Baked Product

Objectives

1. To evaluate the effects of solid versus liquid fat on the quality of a baked product.
2. To gain practice in preparing baked products.

Basic Recipe for Blueberry Muffins

Ingredients

1 ½ c. flour
½ t. salt
½ c sugar
2 t. baking powder
1 egg, lightly beaten
¼ c vegetable oil
½ c. milk
1 c. fresh blueberries or ½ c. canned drained blueberries
Aerosol vegetable oil cooking spray

Procedure

1. Preheat oven to 400 °F. Grease muffin tins with aerosol vegetable oil cooking spray.
2. Sift together the flour, salt, sugar, and baking powder in a mixing bowl.
3. Make a well in the center of the mixture.
4. Combine egg, milk, and oil in a separate bowl.
5. Add the liquid ingredients to the dry ingredients and stir just until moistened. Do not over-mix.
6. Fold berries into batter before the dry ingredients are completely moist.
7. Fill muffin pans $^{2}/_{3}$ full. Bake 20-25 minutes.
8. Evaluate muffins for appearance, flavor, and texture and record observations in Table G-1.

Variations

1. Vegetable oil – Prepare basic recipe for blueberry muffins as above.

2. Shortening – Prepare the recipe as above, except substitute shortening for vegetable oil (do not melt shortening). Blend the fat with the dry ingredients using a pastry blender. Stir in the liquids. The mixture should be lumpy.

Table G-1 Effect of Solid versus Liquid Fat on Quality of a Baked Product			
Fat	**Appearance**	**Flavor**	**Texture**
Vegetable oil			
Shortening			

H. Preparation of Mayonnaise Variations

Objectives

1. To demonstrate the preparation of mayonnaise.
2. To emphasize factors that affect the stability and formation of emulsions.

Note: **Do not eat** any of the mayonnaise variations that contain raw eggs.

Basic Recipe for Mayonnaise

Ingredients

1 egg yolk
1 T. vinegar
¼ t. salt
$^1/_8$ t. dry mustard
$^1/_8$ t. paprika
¼ t. sugar
½ c. vegetable oil

Procedure

1. Put 1 egg yolk into a small bowl.
2. Add seasonings and vinegar.
3. Beat with an electric mixer until well blended.
4. Continue beating at high speed. Add ½ c. oil gradually, by teaspoonfuls and then by tablespoonfuls as the emulsion is formed. Each quantity of oil added should be thoroughly emulsified before the next quantity of oil is added.
5. Evaluate appearance and texture and record observations in Table H-1.

Variations

1. Control, vegetable oil – Prepare basic recipe for mayonnaise as above.
2. Melted butter – Prepare basic recipe for mayonnaise, except substitute melted butter for vegetable oil.
3. Egg white – Prepare basic recipe for mayonnaise, except use an egg white instead of the egg yolk. Leave out dry mustard and paprika.
4. 1 c. vegetable oil – Prepare basic recipe for mayonnaise, except increase vegetable oil from ½ c. to 1 c.
5. Frozen mayonnaise – Prepare basic recipe for mayonnaise. Cover and freeze the mayonnaise. Thaw and observe during the next laboratory period.

6. Blender mayonnaise – Prepare basic recipe for mayonnaise, but use a blender to mix ingredients instead of a rotary beater. Pour in oil in a **slow** stream and continue blending at low speed until all of the oil has been incorporated.

7. Blender mayonnaise, whole egg – Prepare basic recipe for mayonnaise, except substitute one whole egg for the egg yolk and use a blender to mix ingredients instead of a rotary beater.

Table H-1 Preparation of Mayonnaise Variations

Mayonnaise variations	Observations
Control	
Butter	
Egg white	
1 c. oil	
Frozen	
Blender	
Blender – whole egg	

I. Repairing a "Broken" Emulsion

Objectives
1. To observe a "broken" emulsion and become familiar with its characteristics.
2. To demonstrate the technique of reforming an emulsion from a "broken" emulsion.

Basic Procedure to Repair a "Broken" Emulsion

Ingredients
Egg white mayonnaise (variation 3 from Part H)
1 egg yolk

Procedure
1. Use the mayonnaise variation 3 from Part H that was prepared using an egg white. This should have resulted in a "broken" emulsion.
2. Break an egg yolk into a bowl and beat it slightly.
3. Add the broken emulsion slowly, a small amount at a time, to the beaten egg yolk.
4. Beat thoroughly after each addition until emulsification is complete.
5. Evaluate appearance and texture and record observations in Table I-1.

Table I-1 Repairing a "Broken" Emulsion	
	Observations
Broken emulsion	
Addition of egg yolk	

J. Evaluation of Oxidative Rancidity

Basic Procedure for Evaluation of Oxidative Rancidity

Ingredients

Rancid* sunflower oil
Rancid sesame oil
Rancid peanut oil
Rancid vegetable oil
Rancid olive oil

Procedure

1. Keep oils in original containers for display.
2. Evaluate smell and appearance for the presence of oxidative rancidity.
3. Record observations in Table J-1.

* Rancid oils are the result of storing oils in a warm environment for several months.

Table J-1 Evaluation of Oxidative Rancidity		
Product	**Smell**	**Appearance**
Sunflower oil		
Sesame oil		
Peanut oil		
Vegetable oil		
Olive oil		

Additional Food Preparation Exercises

K. Fritters

Objectives

1. To prepare and evaluate food products that are deep-fat fried.
2. To become familiar with factors that influence fat absorption by deep-fried products.

Fritters

Ingredients
1 c. flour
1 ½ t. baking powder
½ t. salt
1 T. granulated sugar (use only in fruit fritters)
1 egg
½ c. milk
1 T. melted butter
½ c. of **one** of the following:
Drained, canned fruit (peaches, pears, fruit cocktail, crushed pineapple) **or**
Fresh, chopped fruit (apples, pears, bananas, peaches, blueberries) **or**
Drained, canned or cooked vegetables (corn, carrots, onions)
Oil for deep-fat frying

Procedure
1. Put oil into a deep-fat fryer to the fill line (or about 3" deep).
2. Sift together all the dry ingredients in a large mixing bowl. Set aside.
3. Place the egg in a small bowl. Add the milk and melted butter and blend thoroughly.
4. Add the liquid ingredients to the dry ingredients all at once. Stir just enough to blend.
5. Using a colander, drain fruit or vegetables thoroughly.
6. Add the fruit or vegetable. Blend with only four or five strokes.
7. Heat the oil to 375 °F.
8. Carefully spoon about 1 T. at a time of the fritter batter into the hot oil.
9. Do not fry more than three fritters at one time. Fry two to three minutes until golden brown. Turn and cook until the second side is golden brown.
10. Remove the cooked fritters. Drain well on paper towels. Keep warm by placing on a baking sheet in an oven preheated to 200 °F.

L. Salad Dressings

Objectives
1. To become familiar with the composition of various types of dressings used in salad preparation.
2. To prepare various salad dressing emulsions.

L-a. Easy Thousand Island Dressing

Thousand Island Dressing

Ingredients
1 c. low-fat mayonnaise
½ c. catsup
¼ - ⅓ c. sweet pickle relish

Procedure
1. Combine all ingredients in a container with a lid and mix well.
2. Serve over tossed green salad or chef's salad.

L-b. Curried Sour Cream-Yogurt Dressing

Sour Cream-Yogurt Dressing

Ingredients

½ c. low-fat cultured sour cream
½ c. plain yogurt
¾ t. curry powder
2 T. pineapple juice

Procedure

1. Combine all ingredients and mix well.
2. Serve over fruit salad or chicken salad.

L-c. Waldorf Salad with Cooked Dressing

Waldorf Salad with Cooked Dressing

Ingredients for Salad

3 apples, cored and cubed, but not peeled
¼ c. raisins
1 c. celery, diced
½ c. chopped nuts of your choice
Cooked dressing (below)

Procedure

1. Lightly mix ingredients together.
2. Add cooked dressing and mix well.

Cooked Dressing

Ingredients for Dressing

2 egg yolks
$^{2}/_{3}$ c. milk
3 T. fresh lemon juice or vinegar
1 T. butter
1 ½ T. sugar
¼ t. salt
2 T. flour
½ t. dry mustard

Procedure

1. Mix together dry ingredients in the top of a double boiler.
2. Slightly beat egg yolks in a small bowl. Add egg yolk and milk to dry ingredients.
3. Place double boiler with mixture over boiling water and cook until thick, stirring continuously.
4. Remove from heat.
5. Add the butter and stir until melted. Add lemon juice.
6. Chill and pour over Waldorf salad.

Post-Lab Study Questions

Textbook reference: Chapters 4 and 22

Discussion Questions

1. Compare the spreadability of regular butter and margarine with that of their whipped counterparts.

2. What was the effect of the coating on the appearance, flavor, and mouthfeel of the fried zucchini?

3. Why do you think the zucchini was sliced in circles rather than in small cubes?

4. Why are both beaten egg and flour used to coat the zucchini rather than just flour alone?

5. Why did the fruit or vegetables need to be well-drained before being mixed into the fritter batter?

6. What is the difference between monounsaturated and polyunsaturated oils?

7. When would you choose to use a flavored oil such as peanut oil instead of vegetable oil?

8. What type of an emulsion is mayonnaise?

9. What are the roles of the dry ingredients in mayonnaise?

10. Why must the oil be added slowly?

11. Why is it not recommended to eat any of the mayonnaise variations that contain raw eggs?

12. Name some factors that can cause an emulsion to “break.”

13. How can a broken oil-in-water emulsion be reestablished?

14. Why is there little danger of breaking the cooked dressing emulsion as it is heated?

15. What ingredient(s) contribute(s) fat to the cooked dressing?

Questions for Post-Lab Writing

16. Compare the kilocalorie content of 1 T. shortening, 1 T. butter, and 1 T. regular margarine.

17. How is the spreadability of a commercial spread related to the composition of the fat? To the temperature of the fat?

18. Based on your experience in the lab, what is the difference between stir-frying and oven frying? How would you expect these methods to affect the fat content as compared with deep-fat frying?

19. Which oil imparted the most pleasing flavor to the cucumber-tomato salad?

20. Name oils that impart a distinctive flavor to foods.

21. What were the differences in appearance, flavor, and texture among the brownies made with fat, those made with applesauce, and those made with baby food? What are the reasons for the differences?

22. What other fat substitutes are available to the consumer?

23. What were the differences in appearance, flavor, and texture between the muffins made with shortening and the muffins made with oil? What are the reasons for the differences?

24. What is the difference between an oil-in-water emulsion and a water-in-oil emulsion?

25. What is the role of the egg yolk in mayonnaise?

26. What causes the mayonnaise to thicken?

27. How do mayonnaise, reduced-fat mayonnaise, fat-free mayonnaise, and Miracle Whip® differ in composition? In kilocalories? In price?

28. What was the effect on the mayonnaise preparation when melted butter was used in place of the oil?

29. Compare the mayonnaise prepared with egg yolk and the one prepared with egg white.

30. Compare the blender mayonnaise to the conventional method mayonnaise prepared with whole eggs.

31. Does freezing influence the stability of mayonnaise? Why?

32. Which ingredients in the cooked dressing function as emulsifying agents? As thickening agents?

33. Why does the cooked dressing need to be heated? Why is it okay to eat?

Unit 16 – Cakes

Introduction

Cakes can be divided into two categories, shortened cakes and foam cakes. The basic ingredients in a shortened cake include flour, fat, sugar, eggs, liquid, baking powder, and flavoring. The proportions of these ingredients must be correctly balanced to produce a cake of high quality. The formation of gluten from flour proteins and the gelatinization of wheat starch contribute to cake structure. Sugar acts as a tenderizing agent, adds flavor, and contributes to the volume of a shortened cake batter during the creaming step. Liquid is necessary for starch gelatinization and the development of gluten. Egg proteins play an important role in providing structure and stabilizing the foam of shortened cakes. The lipoproteins of egg yolk also are effective emulsifying agents. Baking powder is commonly used for the production of carbon dioxide in shortened cakes.

Egg foams are essential in the preparation of foam cakes, such as angel food cakes, sponge cakes, and chiffon cakes. They serve as the primary leavening agent. Egg white foams are stabilized by protein coagulation at the gas-liquid interface. The addition of acid and sugar increases foam stability. Angel food cake is prepared from flour, sugar, egg whites, cream of tartar, and flavorings. Sponge cake is made with whole egg, lemon juice, sugar, flour, and flavorings. The egg whites and yolks are usually beaten separately. Chiffon cakes are similar to sponge cakes but also contain oil and a small amount of chemical leavener. Batters prepared with egg foams require careful folding rather than beating so that the foam does not collapse before baking.

Key Terms

chiffon cake – combination of a shortened and foam cake that includes fat from vegetable oil or egg yolk and egg foams. Cake flour and leavening are also ingredients in a chiffon cake.

foam cake – cake prepared without or with a very small amount of fat from egg yolks; leavened with steam and air from beaten egg foam. Examples include angel food cake and sponge cake.

shortened cake – cake prepared with fat; usually leavened with baking soda or baking powder. Examples include pound cake, chocolate cake, butter cake, etc.

whip – to beat rapidly to increase volume by incorporating air.

Pre-Lab Questions

Textbook reference: Chapters 17 and 23

1. Discuss the functions of fat, sugar, and egg in shortened cakes.

2. Can a sugar substitute be used satisfactorily to replace the sugar in a shortened cake? Why?

3. What type of flour produces the highest quality cake? Why?

4. In a foam cake, why are the dry ingredients folded in rather than beaten in?

5. What is the purpose of the acid source—cream of tarter or lemon juice—in a foam cake?

Lab Procedures

A. Effect of Leavening Agent on Shortened Cakes

Objectives

1. To compare volume, appearance, flavor and texture of yellow cakes prepared with various leavening agents.
2. To compare the effectiveness of homemade baking powder with that of commercial baking powders.
3. To demonstrate the relationship between batter pH and baked product characteristics.
4. To demonstrate the conventional method of mixing.

Basic Recipe for Yellow Cake

Ingredients

1 ½ c. sifted cake flour
¾ c. sugar
¼ c. shortening
1 egg
Dry leavening agent (use assigned variation)
½ t. salt
½ c. milk
½ t. vanilla
Aerosol vegetable oil cooking spray (for pan preparation)
Flour (for pan preparation)

Procedure

1. Grease and flour an 8" square baking pan and set aside. Pre-heat oven to 365 °F.
2. Sift the flour, salt, and dry leavening agent together and set aside.
3. Add the vanilla to the milk and set aside.
4. Cream sugar and shortening together for 2 minutes with an electric mixer at medium speed.
5. Add the egg to the creamed mixture and mix for 1 minute with an electric mixer at medium speed.
6. Add $^{1}/_{3}$ of the flour mixture and $^{1}/_{3}$ of the milk mixture. Mix for 1 minute with an electric mixer at medium speed. Repeat two more times.
7. Continue mixing for 2 minutes with an electric mixer at high speed.
8. Measure the pH of the batter and then transfer it to a greased and floured 8" square baking pan.
9. Bake at 365 °F for approximately 30 minutes. Cake is done when it pulls away from the sides of the pan slightly. Doneness may also be tested by insertion of a toothpick in the center (it should come out clean) or gently pressing the surface of the center of the cake with your fingertip. If the cake springs back after pressing, it is ready to remove from the oven.
10. Cool cake in the pan for 10 minutes on a cooling rack. Invert the pan and release the cake onto a cooling rack. Cool 5 more minutes. Use a sharp knife to cut the cake in half neatly, without compressing it. Use a ruler to measure the height of the cake at the center. Cut into pieces for class evaluation.
11. Evaluate volume (height), appearance, texture, and flavor and record observations in Table A-1.

Characteristics of a high-quality shortened cake:

- Appearance: golden brown, slight arc on top
- Interior: small air cells, creamy color, with ~1-2" height
- Tenderness: no resistance to chew, moist, seems to melt in your mouth
- Flavor: mildly sweet

Variations

1. Double-acting baking powder, 1 ¼ t. – Prepare basic recipe for cake using 1 ¼ t. double-acting baking powder as the dry leavening agent.

2. Double-acting baking powder, 2 ½ t. – Prepare basic recipe for cake using 2 ½ t. double-acting baking powder as the dry leavening agent.

3. No leavener – Prepare basic recipe for cake without a leavener.

4. Homemade baking powder – Prepare homemade baking powder by mixing together ½ t. baking soda, 1 ¼ t. cream of tartar, and ¼ t. cornstarch. Prepare basic recipe for cake using 2 t. homemade baking powder as the dry leavening agent.

5. Baking soda, ½ t. – Prepare basic recipe for cake using ½ t. baking soda as the dry leavening agent.

6. Baking soda, 1 t. – Prepare basic recipe for cake using 1 t. baking soda as the dry leavening agent.

7. Baking soda and lemon juice – Prepare basic recipe for cake using ½ t. baking soda as the dry leavening agent. Add 1 T. lemon juice to the milk.

Table A-1 Yellow Cake with Leavening Variations

Variation	pH	Volume	Appearance	Texture	Flavor
1 ¼ t. baking powder					
2 ½ t. baking powder					
No powder					
Homemade powder					
½ t. baking soda					
1 t. baking soda					
Baking soda + lemon juice					

B. *Effect of Sweeteners on Shortened Cakes*

Objectives

1. To demonstrate the effect of sugar on the volume and tenderness of shortened cakes.
2. To observe the effects of other sweeteners on the volume and tenderness of shortened cakes.

Basic Recipe for Shortened Cake

Ingredients

¼ c. shortening
½ c. granulated sugar
1 egg
1 c. cake flour (sift before measuring)
1 t. baking powder
¼ t. salt
⅓ c. milk
½ t. vanilla

Procedure

1. Sift dry ingredients together and set aside.
2. Cream shortening and sugar in a separate bowl using an electric mixer until light and fluffy.
3. Add the egg to the creamed mixture and beat for 1 minute at medium speed.
4. Add ⅓ of the flour mixture to the creamed mixture and beat for 30 seconds. Add vanilla. Mix 1 minute on medium.
5. Add ½ of the milk mixture to the creamed mixture and beat for 30 seconds.
6. Repeat, alternating dry and wet ingredients until all are incorporated into the creamed mixture.
7. Grease and flour an 8" square cake pan.
8. Bake for 20 to 30 minutes at 350 °F or until a toothpick inserted into the center of the cake comes out clean.
9. Cool cake in the pan for 10 minutes on a cooling rack. Invert the pan and release the cake onto a cooling rack. Cool 5 more minutes. Use a sharp knife to cut the cake in half neatly, without compressing it. Use a ruler to measure the height of the cake at the center. Cut into pieces for class evaluation.
10. Evaluate the appearance, volume (height), tenderness, and flavor of the cakes, and record observations in Table B-1.

Variations

1. Control – Follow the basic recipe for a shortened cake above.

2. No granulated sugar – Follow the basic recipe for a shortened cake, except omit granulated sugar.

3. Doubled granulated sugar – Follow the basic recipe for a shortened cake, except double the amount of granulated sugar.

4. Honey – Follow the basic recipe for a shortened cake, except substitute ¼ c. + 2 T. honey for granulated sugar.

5. Sugar substitute – Follow the basic recipe for a shortened cake, except substitute an amount of nonnutritive sweetener equivalent to ½ c. sugar (½ c. Splenda®).

Table B-1 Effect of Sweeteners on Shortened Cakes				
Sweetener source	**Appearance**	**Volume**	**Tenderness**	**Flavor**
Control, ½ c. sugar				
No sugar				
Double sugar				
Honey				
Nonnutritive sweetener				

C. Effect of Fats on Shortened Cakes

Objectives

1. To demonstrate the effect of fat on the volume and tenderness of shortened cakes.
2. To observe the effects of various types of fat on the volume and tenderness of shortened cakes.

Basic Recipe for Shortened Cake

Ingredients

¼ c. shortening
½ c. granulated sugar
1 egg
1 c. cake flour (sift before measuring)
1 t. baking powder
¼ t. salt
⅓ c. milk
½ t. vanilla

Procedure

1. Sift dry ingredients together and set aside.
2. Cream shortening and sugar in a separate bowl using an electric mixer until light and fluffy.
3. Add the egg to the creamed mixture and beat for 1 minute at medium speed.
4. Add ⅓ of the flour mixture to the creamed mixture and beat for 30 seconds.
5. Add the vanilla to the milk. Add ½ of the milk mixture to the creamed mixture and beat for 30 seconds.
6. Repeat, alternating dry and wet ingredients until all are incorporated into the creamed mixture.
7. Grease and flour an 8" square cake pan.
8. Bake for 20 to 30 minutes at 350 °F or until a toothpick inserted into the center of the cake comes out clean.
9. Cool cake in the pan for 10 minutes on a cooling rack. Invert the pan and release the cake onto a cooling rack. Cool 5 more minutes. Use a sharp knife to cut the cake in half neatly, without compressing it. Use a ruler to measure the height of the cake at the center. Cut into pieces for class evaluation.
10. Evaluate the appearance, volume (height), tenderness, and flavor of the cakes, and record observations in Table C-1.

Variations

1. Control – Follow the basic recipe for a shortened cake as above.

2. No shortening – Follow the basic recipe for a shortened cake, except omit shortening.

3. Double shortening – Follow the basic recipe for a shortened cake, except double the amount of shortening.

4. Butter – Follow the basic recipe for a shortened cake, except substitute ¼ c. butter for shortening.

5. Oil – Follow the basic recipe for a shortened cake, except substitute ¼ c. oil for shortening.

Table C-1 Effect of Fats on Shortened Cakes

Fat source	Appearance	Volume	Tenderness	Flavor
Control, ¼ c. shortening				
No shortening				
Double shortening				
Butter				
Oil				

D. Effect of Flour Type on Cake

Objectives
1. To compare the volume, texture, and flavor of cakes prepared using different types of flour.
2. To illustrate the conventional method of mixing.

Basic Recipe for Yellow Cake

Ingredients
¾ c. sugar
¼ c. shortening
1 egg
1 ½ c. flour
1 ¼ t. double-acting baking powder
½ t salt
½ c. milk
½ t. vanilla
Aerosol vegetable oil cooking spray (for pan preparation)
Flour (for pan preparation)

Procedure
1. Grease and flour an 8” square baking pan and set aside. Pre-heat oven to 365 °F.
2. Sift the flour, salt, and baking powder together and set aside. (Do not sift whole-wheat flour—just mix it with salt and baking powder.)
3. Add the vanilla to the milk and set aside.
4. Cream sugar and shortening together for 2 minutes with an electric mixer at medium speed.
5. Add the egg to the creamed mixture and mix for 1 minute with an electric mixer at medium speed.
6. Add $^1/_3$ of the flour mixture and $^1/_3$ of the milk mixture. Mix for 1 minute with an electric mixer at medium speed. Repeat two more times.
7. Continue mixing for 2 minutes with an electric mixer at high speed.
8. Transfer batter to a greased and floured 8” square baking pan.
9. Bake at 365 °F for approximately 30 minutes. Cake is done when it pulls away from the sides of the pan slightly. Doneness may also be tested by insertion of a toothpick in the center (it should come out clean) or gently pressing the surface of the center of the cake with your fingertip. If the cake springs back after pressing, it is ready to remove from the oven.
10. Cool cake in the pan for 10 minutes on a cooling rack. Invert the pan and release the cake onto a cooling rack. Cool 5 more minutes. Use a sharp knife to cut the cake in half neatly, without compressing it. Use a ruler to measure the height of the cake at the center. Cut into pieces for class evaluation.
11. Evaluate volume, texture, and flavor of each cake prepared and record observations in Table D-1.

Variations

1. Cake flour, control – Follow the basic recipe for cake using cake flour.
2. All-purpose flour – Follow the basic recipe for cake using all-purpose flour.
3. Bread flour – Follow the basic recipe for cake using bread flour.
4. Whole-wheat flour – Follow the basic recipe for cake using whole-wheat flour.
5. Rye – Follow the basic recipe for cake using rye flour.
6. Buckwheat – Follow the basic recipe for cake using buckwheat flour.

Table D-1 Comparison of Flour Types

Flour	Volume (height)	Flavor	Texture
Cake			
All-purpose			
Bread			
Whole-wheat			
Rye			
Buckwheat			

E. Comparison of Foam Cakes

Objectives

1. To compare various types of foam cakes.
2. To demonstrate the preparation of an angel food cake.
3. To demonstrate the preparation of a sponge cake.
4. To demonstrate the preparation of a chiffon cake.
5. To emphasize the functions of the ingredients in foam cakes.

E-a. Basic Angel Food Cake

Basic Recipe for Angel Food Cake

Ingredients

1 c. cake flour
1 ½ c. sugar
1 ½ c. room temperature egg whites (~10 large eggs)
1 ½ t. vanilla extract
1 ½ t. cream of tarter
½ t. salt

Procedure

1. Sift ¾ c. of the sugar and the flour together twice and set aside.
2. Beat the egg whites until foamy.
3. Add the cream of tartar and salt and beat to the soft peak stage.
4. Add the rest of the sugar, a tablespoon at a time, beating thoroughly after each addition.
5. Continue beating to the stiff peak stage.
6. Sift the flour mixture, about one fourth at a time, over the egg whites, folding the mixture in gently with a rubber spatula after each addition using an over-under motion. Avoid over-folding the mixture but be sure all of the dry ingredients have been incorporated.
7. Pour the batter into an ungreased tube pan.
8. Bake at 350 °F for 40 to 50 minutes. Invert the baking pan and allow the cake to cool before removing it from the pan. Measure the cake's height in centimeters.
9. Evaluate the appearance, aroma, flavor, and texture, and record in Table E-1.

E-b. Basic Sponge Cake

Basic Recipe for Sponge Cake

Ingredients

1 c. cake flour, sifted before measuring
1 c. granulated sugar
¼ t. salt
½ t. cream of tartar
5 large eggs, separated
3 T. cold water
½ t. vanilla or lemon extract

Procedure

1. Add the flour to one third of the sugar and sift. Set aside.
2. Beat the egg yolks with the water and flavoring until the mixture is very thick and light yellow.
3. Add one third of the sugar to the egg yolk mixture and beat thoroughly.
4. Beat the egg whites until foamy. Add salt and cream of tartar and beat to the soft peak stage.
5. Sprinkle the remaining sugar over the whites and beat until the meringue reaches the stiff peak stage.
6. Sift one-fourth of the flour-sugar mixture over the egg-yolk mixture and **gently** fold it in with a rubber spatula, using an over-under motion. Repeat until all of the flour-sugar mixture has been incorporated.
7. Combine the egg white mixture and the egg yolk mixture. Fold **gently** until blended.
8. Bake in an ungreased tube pan at 350 °F for 40 to 45 minutes.
9. Invert the baking pan and allow the cake to cool before removing it from the pan. Measure height.
10. Evaluate the appearance, volume, tenderness, and flavor of the cake, and record in Table E-1.

E-c. Basic Chiffon Cake

Basic Recipe for Chiffon Cake

Ingredients

2 ¼ c. cake flour, sifted before measuring
1 ½ c. granulated sugar
3 t. baking powder
½ t. salt
¾ c. cold water
½ c. oil
5 egg yolks
1 t. vanilla or lemon extract
1 c. egg whites, room temperature
½ t. cream of tartar

Procedure

1. Combine flour, 1 ½ c. sugar, baking powder, and salt and sift. Set aside.
2. Beat the oil and egg yolk together. Stir in the water and the flavoring.
3. Combine the liquid and dry ingredients and mix just until smooth.
4. Beat the egg whites to the foamy stage. Add the cream of tartar and continue beating to the soft peak stage.
5. Fold the batter mixture into the egg white foam with a rubber spatula, using an over-and-under motion.
6. Pour the batter into an ungreased tube pan and bake at 350 °F for 50-60 minutes or until cake springs back when touched with fingertip.
7. Invert the baking pan and allow the cake to cool before removing it from the pan. Measure height.
8. Evaluate the appearance, volume, tenderness, and flavor of the cake, and record in Table E-1.

Table E-1 Comparison of Foam Cakes				
Foam cakes	**Appearance**	**Volume (height)**	**Tenderness**	**Flavor**
Angel food				
Sponge				
Chiffon				

Additional Food Preparation Exercises

F. Preparation of Selected Cakes

F-a. Grandma's Gingerbread

Grandma's Gingerbread

Ingredients

½ c. butter
½ c. granulated sugar
1 egg, well beaten
1 c. molasses
2 ½ c. flour, sifted
1 c. hot water
1 t. salt
1 t. ground cloves
1 t. ginger
1 t. cinnamon
1 ½ t. baking soda
Aerosol vegetable oil cooking spray

Procedure

1. Cream butter and sugar in a large mixing bowl. Add beaten egg and molasses and mix well.
2. Sift dry ingredients together, including spices, and add to creamed mixture.
3. Add hot water and beat until smooth.
4. Spray 9" × 9" pan with aerosol vegetable oil cooking spray. Pour batter into prepared pan and bake at 375 °F for 35 to 40 minutes or until toothpick inserted in center comes out clean.
5. Let cool approximately 5 minutes before transferring cake from baking pan to cooling rack.

From the recipe file of Alma Mynhier

F-b. Banana Nut Cake

Banana Nut Cake

Cake Ingredients

1 stick margarine
1 ½ c. sugar
2 eggs
2 c. + 2 T. flour (divided)
2 t. soda
Pinch of salt
6 T. buttermilk
2 ripe bananas, mashed
1 c. chopped pecans
1 t. vanilla
Shortening

Procedure

1. Preheat oven to 375 °F.
2. Cream margarine and sugar in a large mixing bowl. Add eggs to the creamed mixture and beat for 1 minute at medium speed.
3. Sift dry ingredients together.
4. Add $^{1}/_{3}$ of the flour mixture to the creamed mixture and beat for 30 seconds.
5. Add ½ of the buttermilk to the creamed mixture and beat for 30 seconds.
6. Repeat, alternating dry and wet ingredients until all are incorporated into the creamed mixture.
7. Mix in bananas, pecans, and vanilla.
8. Grease and flour three 9" round cake pans. Pour batter into prepared pans and bake at 375 °F for 30 minutes or until toothpick inserted in center comes out clean.
9. Transfer baked cakes to cooling racks.
10. When completely cool, place 1 round cake on cake plate and ice it using the recipe below. Stack second round cake on top of first and ice it. Do the same with the third round cake.

Icing Ingredients

¾ c. shortening
¼ c. margarine
½ t. salt
1 t. cream of tarter
1 t. vanilla
¼ c. water
1 box powdered sugar

Procedure

1. Cream shortening and margarine in a large mixing bowl.
2. Sift powdered sugar with salt and cream of tarter; add dry ingredients to the creamed mixture.
3. Blend in water and vanilla and beat 10 minutes at medium speed. Ice cake according to above procedure.

From the recipe file of Alma Mynhier

Post-Lab Study Questions

Textbook reference: Chapters 12, 17, 21, and 23

Discussion Questions

1. Which leavening systems caused the greatest increase in volume? How would you explain this?

2. Which leavening system caused the least increase in volume? How would you explain this?

3. Compare the cakes prepared with no sugar and double sugar to the cake prepared from the standard recipe.

4. Compare the cakes prepared with no shortening and double shortening to the cake prepared from the standard recipe.

5. What is the advantage of using a plastic fat in a shortened cake rather than oil?

6. Which would you expect to produce a cake with the most desirable texture—butter or shortening? Why?

7. What characteristics of the cake were affected when different flours were used?

8. Why was only $^{3}/_{8}$ c. honey substituted for the ½ c. sugar in the shortened cake?

Questions for Post-Lab Writing

9. What ingredients must be present in a cake to cause adequate leavening if baking soda is used?

10. Why do you think cake flour makes superior cakes? If the flour is not sifted before measuring, how might the product be affected?

11. What characteristics possessed by cake flour help create the qualities desired in a cake?

12. Why is part of the sugar in a foam cake incorporated into the egg-white foam?

13. Why are foam cakes cooled in the inverted position?

Unit 17 – Pastry

Introduction

The term *pastry* may be used in a broad sense to include a variety of products made from stiff dough that is relatively high in fat, such as sweet rolls and puff pastry. However, the term is more commonly used to mean plain pastry or pie crust. Pastry ingredients include fat, flour, liquid, and salt. Both flakiness and tenderness are desirable characteristics of pastry. Tenderness results when the fat coats the flour in such a way that hydration of the flour particles is minimized. Pastry tenderness can be evaluated objectively with the Bailey shortometer. This instrument measures the force (breaking strength) required to break small wafers of plain pastry, crackers, or cookies. A greater amount of force is required to break a less tender product. The firmness of fat and its distribution directly affect pastry flakiness. Using cold fat minimizes its absorption by the flour, creating pea-size balls of fat. These pea-size balls are dispersed throughout the flour and form pockets of air when heated, resulting in flakiness.

Many factors influence pastry quality, including the extent of manipulation and the amounts and types of ingredients. The protein content of the flour used influences the amount of gluten formed and therefore also influences pastry tenderness. Plasticity of the fat used is a factor in determining both tenderness and flakiness of the pastry. Plastic fats such as hydrogenated shortening and lard result in the flakiest crusts. Adding excess water, or increasing the amount of manipulation after water is added, causes extensive hydration of the flour proteins. This results in increased gluten production and a less tender pastry. However, insufficient water produces dry, crumbly dough that is difficult to roll. Lack of water does not allow enough steam to form to produce the flaky layers characteristic of a good pastry.

Key Terms

Bailey shortometer – instrument used to objectively evaluate the tenderness of pastry, crackers, or cookies by measuring the amount of force required to break a product.

blind bake – baking an unfilled pie crust.

cut in – to distribute solid fat in dry ingredients with knives, with a pastry blender, or in a mixer with proper attachments until fat particles are desired size.

lamination – alternating layers of fat and flour in pastry dough. Fat melts when the pastry is baked and leaves empty spaces where steam lifts the layers of flour and produces a flakey texture.

mealy – describes grainy pastry that results from completely coating flour with fat.

pastry guides – tools used to control the thickness of a pie pastry when rolling it out.

puff pastry – pastry with multiple, alternating layers of fat and flour that puffs when baked. After baking, puff pastry may be filled with various fillings.

strudel – German pastry prepared with a layer of thin dough filled with fruit or cream cheese. The pastry becomes very flakey when baked.

Pre-Lab Questions

Textbook reference: Chapter 24

1. What is the purpose of cutting the shortening into the flour?

2. What fat source is recommended to produce the best pie pastry? Why?

3. How does the amount of manipulation after water is added affect the final pastry product?

4. Why is chilling the dough recommended for pastry preparation?

5. What are the characteristics of a quality pastry?

Figure 17-1: Correct use of a pastry blender.

Lab Procedures

A. *Effect of Type of Fat*

Objectives

1. To demonstrate the effect of plasticity of fat on pastry tenderness and flakiness.
2. To gain practice in pastry preparation.

Basic Recipe for Pie Pastry

Ingredients

½ c. all-purpose flour
¼ t. salt
2 T. refrigerated fat
1 ½ T. cold water

Procedure

1. Sift flour and salt together into a mixing bowl.
2. Cut the fat into the flour with a pastry blender or a knife until the mixture resembles cornmeal (see Figure 17-1).
3. Using ½ t. at a time, sprinkle the water over the mixture. Lightly toss the moistened mixture on top to the side of the bowl with a fork before adding more water.
4. When all of the water has been added, gather the mixture into a ball and press together gently. If dough seems dry, do not add more water. Check with the instructor.
5. Place the dough on a piece of waxed paper and gently flatten it out with your fingers. Place pastry guides on both sides of the dough. Place another piece of waxed paper on top of the dough and roll the dough until it is the same thickness as the guides ($^1/_8$").
6. Cut the dough into 1" × 5" rectangles. Loosen pastry dough carefully from the waxed paper. Do not stretch the pastry.
7. Place rectangles on an ungreased baking sheet and bake in a 425 °F oven for 10 minutes.
8. Measure the height of 5 pastry wafers stacked on top of each other; this is an indirect way to measure flakiness.
9. Use half of the total number of rectangles to measure breaking strength of pastry rectangles using a shortometer or other texture analysis equipment, if available (see Figure 17-2).
10. Using the remaining rectangles, evaluate the flavor, tenderness, and flakiness of the pastry and record observations in Table A-1.

Characteristics of a high-quality pastry:

- Appearance: opaque, creamy with elevated areas golden brown; slightly uneven surface (air pockets cause puffing of surface).
- Texture: layers of flakes are visible with a side view. Size of flakes depends on crystal size of solid fat. Not mealy.
- Mouthfeel: crisp, flaky.
- Tenderness: slight resistance to chew, easily cut with a fork.

Figure 17-2: Measuring a muffin's properties with a texture analyzer.

Variations

1. Shortening – Prepare the basic recipe for pie pastry as above. Use refrigerated shortening as the fat source.

2. Oil – Prepare the basic recipe for pie pastry, except use refrigerated vegetable oil as the fat source. Stir in the oil with a fork.

3. Butter – Prepare the basic recipe for pie pastry, except use refrigerated butter as the fat source.

4. Margarine – Prepare the basic recipe for pie pastry, except use refrigerated margarine as the fat source.

5. Lard – Prepare the basic recipe for pie pastry, except use refrigerated lard as the fat source.

6. Lard, coarse pieces – Prepare the basic recipe for pie pastry, except use refrigerated lard as the fat source and cut it into the flour, using 2 knives, only until clearly visible ~$^{1}/_{8}$" pieces are formed.

Table A-1 Effect of Type of Fat					
Type of fat	**Breaking strength**	**Stacked height**	**Flavor**	**Tenderness**	**Flakiness**
Shortening					
Oil					
Butter					
Margarine					
Lard					
Lard, coarse pieces					

B. Effect of Type of Flour

Objectives

1. To demonstrate the effect of type of flour on pastry tenderness.
2. To gain practice in pastry preparation.

Basic Recipe for Pie Pastry

Ingredients

½ c. flour
¼ t. salt
2 T. shortening
1 ½ T. cold water

Procedure

1. Sift flour and salt together into mixing bowl.
2. Cut the fat into the flour with a pastry blender or a knife until the mixture resembles cornmeal.
3. Using ½ t. at a time, sprinkle the water over the mixture. Lightly toss the moistened mixture on top to the side of the bowl with a fork before adding more water.
4. When all of the water has been added, gather the mixture into a ball and press together gently. If dough seems dry, do not add more water. Check with the instructor.
5. Place the dough on a piece of waxed paper and gently flatten it out with your fingers. Place pastry guides on both sides of the dough. Place another piece of waxed paper on top of the dough and roll the dough until it is the same thickness as the guides ($^1/_8$").
6. Cut the dough into 1" × 5" rectangles. Loosen pastry dough carefully from the waxed paper. Do not stretch the pastry.
7. Place rectangles on an ungreased baking sheet and bake in a 425 ºF oven for 10 minutes.
8. Measure the height of 5 pastry wafers stacked on top of each other; this is an indirect way to measure flakiness.
9. Use half of the total number of rectangles to measure breaking strength of pastry rectangles using a shortometer or other texture analysis equipment, if available.
10. Using the remaining rectangles, evaluate the flavor, tenderness, and flakiness of the pastry and record observations in Table B-1.

Variations

1. All-purpose flour – Prepare the basic recipe for pie pastry as above using all-purpose flour.

2. Pastry – Prepare the basic recipe for pie pastry as above using pastry flour.

3. Whole-wheat flour – Prepare the basic recipe for pie pastry as above using whole-wheat flour.

4. Cake flour – Prepare the basic recipe for pie pastry as above using cake flour.

Table B-1 Effect of Type of Flour

Type of flour	Breaking strength	Stacked height	Flavor	Tenderness	Flakiness
All-purpose					
Pastry					
Whole-wheat					
Cake					

C. *Effect of Extent of Manipulation*

Objectives
1. To demonstrate the effect of manipulation on pastry tenderness and flakiness.
2. To gain practice in pastry preparation.

Basic Recipe for Pie Pastry

Ingredients
½ c. flour
¼ t. salt
2 T. shortening
1 ½ T. cold water

Procedure
1. Sift flour and salt together into small mixing bowl.
2. Cut the fat into the flour with a pastry blender or a knife until the mixture resembles cornmeal.
3. Using ½ t. at a time, sprinkle the water over the mixture. Lightly toss the moistened mixture on top to the side of the bowl with a fork before adding more water.
4. When all of the water has been added, gather the mixture into a ball and press together gently. If dough seems dry, do not add more water. Check with the instructor.
5. Place the dough on a piece of waxed paper and gently flatten it out with your fingers. Place pastry guides on both sides of the dough. Place another piece of waxed paper on top of the dough and roll the dough until it is the same thickness as the guides ($^1/_8$").
6. Cut the dough into 1" × 5" rectangles. Loosen pastry dough carefully from the waxed paper. Do not stretch the pastry.
7. Place rectangles on an ungreased baking sheet and bake in a 425 °F oven for 10 minutes.
8. Measure the height of 5 pastry wafers stacked on top of each other; this is an indirect way to measure flakiness.
9. Use half of the total number of rectangles to measure breaking strength of pastry rectangles using a shortometer or other texture analysis equipment, if available.
10. Using the remaining rectangles, evaluate the flavor, tenderness, and flakiness of the pastry and record observations in Table C-1.

Variation

1. No knead – Prepare basic recipe for pie pastry as above.
2. Knead 2 minutes – Prepare basic recipe for pie pastry as above, except knead the pastry dough for 2 minutes after gathering the mixture into a ball (step 4).
3. Knead 4 minutes – Prepare basic recipe for pie pastry as above, except knead the pastry dough for 4 minutes after gathering the mixture into a ball (step 4).

Table C-1 Effect of Extent of Manipulation					
Kneading time	**Breaking strength**	**Stacked height**	**Flavor**	**Tenderness**	**Flakiness**
0 minutes					
2 minutes					
4 minutes					

D. Comparison of Commercial Pie Pastries

Objectives

1. To compare the ease of preparation of selected commercial pie crusts and mixes.
2. To compare the flavor, tenderness, and flakiness of selected commercial pie crusts and mixes.

Basic Procedure for Commercial Pie Crust Preparation

Ingredients

Commercial pie crust (see variations)

Procedure

1. Prepare pie crust according to commercial package directions.
2. Cut the dough into 1" × 5" rectangles. Bake according to package directions.
3. Measure the height of 5 pastry wafers stacked on top of each other; this is an indirect way to measure flakiness.
4. Use half of the total number of rectangles to measure breaking strength of pastry rectangles using a shortometer or other texture analysis equipment, if available.
5. Using the remaining rectangles, evaluate the flavor, tenderness, and flakiness of the pastry and record observations in Table D-1.

Variations

1. Frozen ready-to-bake pie crust – Prepare frozen ready-to-bake pie crust according to above procedure.

2. Stick-type pie crust mix – Prepare stick-type pie crust according to above procedure.

3. Dry pie crust mix – Prepare dry pie crust mix according to above procedure.

Table D-1 Comparison of Commercial Pie Pastry					
Commercial pastry	**Breaking strength**	**Stacked height**	**Flavor**	**Tenderness**	**Flakiness**
Frozen, ready-to-bake					
Stick-type mix					
Dry mix					

Additional Food Preparation Exercises

E. Preparation of Various Pies

Objectives
1. To gain experience in pastry preparation.

E-a. Turkey Pot Pie

Turkey Pot Pie

Ingredients for Double Crust
2 c. flour
½ t. salt
1 c. Crisco® shortening
6 T. water

Ingredients for Filling
1/3 c. margarine
1/3 c. onions, chopped
1/3 c. flour
½ t. salt
¼ t. pepper
1 c. chicken broth
2/3 c. milk
3 c. shredded turkey or chicken
1 c. frozen mixed vegetables, thawed
2 c. potatoes, chopped

Procedure
1. Preheat oven to 425 ºF.
2. Sift flour and salt together into small mixing bowl.
3. Cut the refrigerated fat into the flour with a pastry blender or a knife until the mixture resembles cornmeal.
4. Using ½ t. at a time, sprinkle the water over the mixture. Lightly toss the moistened mixture on top to the side of the bowl with a fork before adding more water.
5. When all of the water has been added, gather the mixture into a ball and press together gently. If dough seems dry, do not add more water. Check with the instructor. Refrigerate dough for 30 minutes.
6. Remove half of the dough from the refrigerator and place it on a piece of waxed paper. Gently flatten it out with your fingers. Place another piece of waxed paper on top of the dough and roll the dough until it is the desirable thickness—1/8" and 10" diameter. Transfer to 9" pie pan.
7. In a medium saucepan, melt margarine over medium heat. Add onions and celery and cook 2 minutes or until tender. Stir in flour, salt, and pepper until well blended.
8. Gradually stir in broth and milk; cook, stirring constantly, until bubbly and thickened.
9. Add chicken or turkey, potatoes, and mixed vegetables; remove from heat. Spoon mixture into crust-lined pan.
10. Remove remaining pastry dough from the refrigerator and place it on a piece of waxed paper. Gently flatten it out with your fingers. Place pastry guides on both sides of the dough. Place another piece of waxed paper on top of the dough and roll the dough until it is the desirable thickness—1/8" and 10" diameter.
11. Top turkey mixture with second crust and flute; cut slits in several places.
12. Bake for 30-40 minutes or until crust is brown. Let stand 2 minutes before serving.

From the recipe file of Karen Beathard

E-b. Lemon Pie

Lemon Pie

Ingredients for Crust

1 c. flour
½ t. salt
¼ c. shortening
3 T. water

Ingredients for Filling

3 egg yolks
¾ c. sugar
3 T. cornstarch (level)
¼ c. margarine, melted
¼ c. real lemon juice
1 c. whole milk
1 c. sour cream

Ingredients for Meringue

3 egg whites
6 T. granulated sugar

Procedure

1. Preheat oven to 425 ºF.
2. Sift flour and salt together into small mixing bowl.
3. Cut the refrigerated fat into the flour with a pastry blender or a knife until the mixture resembles cornmeal.
4. Using ½ t. at a time, sprinkle the water over the mixture. Lightly toss the moistened mixture on top to the side of the bowl with a fork before adding more water.
5. When all of the water has been added, gather the mixture into a ball and press together gently. If dough seems dry, do not add more water. Check with the instructor. Refrigerate dough for 30 minutes.
6. Remove dough from the refrigerator and place it on a piece of waxed paper. Gently flatten it out with your fingers. Place another piece of waxed paper on top of the dough and roll the dough until it is the desirable thickness—$^{1}/_{8}$" and 10" diameter. Transfer to 9" pie pan and bake approximately 10 minutes or until golden brown.
7. Once crust is baked, reduce oven temperature to 350 ºF.
8. Separate egg yolks from whites. Set egg whites aside.
9. Beat egg yolks in a mixing bowl. Add sugar, cornstarch, margarine, milk, and lemon juice. Mix well.
10. Transfer mixture to a saucepan and cook until very thick, stirring constantly.
11. Cool. Fold in sour cream. Pour into cooked crust.
12. Beat egg whites with rotary beater until they just reach the soft peak stage.
13. Gradually add 6 T. sugar and beat slowly until the meringue is very thick but will still form peaks that have rounded tops. Do not beat to the stiff peak stage.
14. Transfer meringue to pie and bake at 350 ºF for 10 minutes.

From the recipe file of Alma Mynhier

E-c. Pecan Pie

Pecan Pie

Ingredients for Crust

1 c. flour
½ t. salt
¾ c shortening
3 T. water

Ingredients for Filling

3 eggs, beaten
1 c. sugar
1 c. Karo® syrup
1 c. pecan halves
¼ t. vanilla extract

Procedure

1. Sift flour and salt together into small mixing bowl.
2. Cut the refrigerated fat into the flour with a pastry blender or a knife until the mixture resembles cornmeal.
3. Using ½ t. at a time, sprinkle the water over the mixture. Lightly toss the moistened mixture on top to the side of the bowl with a fork before adding more water.
4. When all of the water has been added, gather the mixture into a ball and press together gently. If dough seems dry, do not add more water. Check with the instructor. Refrigerate dough for 30 minutes.
8. Preheat oven to 325 °F.
9. Remove dough from the refrigerator and place it on a piece of waxed paper. Gently flatten it out with your fingers. Place another piece of waxed paper on top of the dough and roll the dough until it is the desirable thickness—$^{1}/_{8}$" and 10" diameter. Transfer pastry to 9" pie pan and set aside.
10. Combine Karo® syrup and sugar in a saucepan. Warm over low heat, stirring constantly, until sugar dissolves.
11. Let cool slightly, then add beaten eggs and vanilla extract. Mix well.
12. Pour mixture into unbaked pastry shell and top with pecan halves.
13. Bake at 325 °F for 10 minutes. Reduce temperature to 300 °F and cook an additional 50 minutes.

From the recipe file of Laurie Lindsey

Post-Lab Study Questions

Textbook reference: Chapter 24

Discussion Questions

1. How is a pastry affected if the fat particles in the mixture are too large?

2. Describe the desired structure of a baked pastry.

3. Do the objective measurements of tenderness correlate with the sensory evaluation of tenderness? If not, give some possible reasons for the inconsistencies.

4. Compare the commercial pie pastries with the basic pastry (all-purpose flour, 2 tablespoons shortening) made from "scratch." Which is most convenient? Most expensive? Most palatable?

Questions for Post-Lab Writing

5. Are there apparent differences in the tenderness of pastries prepared with different fats? Explain any differences.

6. Explain the relationship between plasticity of a fat and pastry tenderness and flakiness.

7. Compare the pastries prepared with different types of flour. Explain any differences.

8. Explain the effect of kneading on pastry tenderness and flakiness.

9. Does chilling the dough influence the final quality of a pastry? If so, in what way?

10. Were the breaking strength measurements more or less consistent among sample of a ready-to-bake pastry shell as compared to hand-rolled pastries? If there was a difference, what might explain this?

Unit 18 – Candy

Introduction

Candies may be classified as crystalline (e.g., fondant and fudge) or noncrystalline (amorphous, e.g., caramels, marshmallows, taffy, brittles, and toffee). The characteristics of crystalline candy are influenced by the ingredients, the rate of cooking, the final concentration of sugar, the conditions during cooling, and the degree of agitation during crystallization. Steps in the preparation of crystalline candies include: 1) dissolving the sugar; 2) concentrating the solution; 3) supersaturating the solution; and 4) controlled crystallization resulting in the desired size of sugar crystals. The sugar must be completely dissolved initially so that no crystals will be present to seed the concentrated sugar solution and start premature crystallization. The boiling point of water is increased by dissolved sugar. Since the boiling point of a sugar solution is directly related to the number of moles of sugar present, the final boiling temperature of a candy mixture is commonly used as an indicator for doneness. An alternate objective test of the appropriate concentration of the sugar solution is the cold water test.

A solution is supersaturated when the amount of sugar in the solution exceeds the amount theoretically possible at a given temperature. Carefully cooling the concentrated sugar solution, without any jostling or agitation, results in the formation of such a supersaturated solution. Once supersaturated, the mixture is rapidly agitated to promote crystallization. This technique results in a candy with a smooth, creamy texture. Adequate cooling followed by continuous, rapid agitation is essential for the formation of small sucrose crystals.

Interfering agents, including sugars other than sucrose, fats, milk solids, and acids, aid in keeping sucrose crystals small. Acids promote hydrolysis, converting some of the sucrose molecules to glucose and fructose, which interfere with the formation of sucrose crystals. Fats, if present, inhibit crystal growth by coating small sucrose crystals as they form, making it difficult for more sucrose molecules to attach to the existing crystals. Milk solids increase the viscosity of the candy mixture. This inhibits the movement of sucrose molecules through the mixture, making it more difficult for them to reach and bind to existing sucrose crystals.

Crystallization is prevented in noncrystalline candies by adding a comparatively large quantity of interfering agents, by allowing no beating or agitation to form nuclei for crystals, and by concentrating the mixture to a high degree so that the mixture will be extremely viscous and will solidify rapidly, preventing molecules from moving into ordered crystals.

The complex changes that occur during the heating of dry sucrose are referred to as caramelization. In this process, sucrose is decomposed into dehydrated monosaccharides and their polymers. Organic sugar acids also are produced. These substances contribute characteristic flavors and color to candies that are cooked to high temperatures.

Key Terms

candy (as in, "to candy") – to cook in heavy syrup to preserve or glaze.

caramelize – to heat sugar, or foods containing sugar, until the sugar melts and a brown color and characteristic flavor develop.

cold water test – objective test used to determine a syrup's consistency. A small amount of syrup is placed in a cup of cold water and its softness or firmness is assessed.

crystalline candy – soft, smooth, creamy candy made of small sugar crystals. Examples of crystalline candy include fudge, fondant, pralines, and divinity.

fondant – creamy, white crystalline candy that is used as icing to glaze and decorate various products and as the filling of many candies including chocolates.

interfering agents – ingredients added to the candy syrup that prevent sugar molecules from clustering together and forming large crystals.

noncrystalline candy – candy without organized structure or form; no crystals are present. Examples include caramel, peanut brittle, taffy, hard candy, and gummy candy.

Pre-Lab Questions

Textbook reference: Chapters 21 and 25

1. What is a "supersaturated" solution? How is it prepared? Why is it important in candy preparation?

2. What is the purpose of the baking soda in peanut brittle?

3. Explain how to perform the cold water test. What information does the cold water test provide?

4. Describe how to calibrate a candy thermometer.

5. How does the temperature at which the beating of a crystalline candy is initiated influence the quality of the final product?

Lab Procedures

A. Objective Testing Methods

Objectives

1. To become familiar with the different objective testing methods.
2. To learn the proper method of thermometer calibration.
3. To become familiar with the cold water test.
4. To become familiar with the texture of syrup cooked to various temperatures after cold water exposure.

A-a. Thermometer Calibration

Basic Procedure for Thermometer Calibration

Ingredients

2 c. water

Procedure

1. Place water in a sauce pan and bring to a boil.
2. Place the candy thermometer in the boiling water and allow the temperature to register. The thermometer should register 212 °F, the correct boiling point for water.
3. If the thermometer does not register the correct boiling point for water, adjust the endpoint temperatures used for boiled sugar syrups.
4. Calibrate all candy thermometers before using.

A-b. Cold Water Test for Sugar Syrup

Basic Procedure for Cold Water Test

Ingredients

1 c. sugar

½ c. water + additional water as needed

Ice

Procedure

1. Mix 1 c. sugar with ½ c. of water.
2. Position a candy thermometer so that the bulb is immersed in the mixture.
3. After heating the sugar syrup to 234 °F to 240 °F (112 °C to 116 °C), drop ½ t. of hot syrup into a clear cup of ice water.
4. Observe and record the appearance, color, and texture of the syrup in the water in Table A-1.
5. Compare your observations to the stages listed in Table A-2, "Cold Water Test for Sugar Syrups."

Variations

1. 234 °F to 240 °F – Follow basic procedure for the cold water test as above.

2. 244 °F to 248 °F – Follow basic procedure for the cold water test, but continue to heat syrup to 244 °F to 248 °F (118-120 °C) before dropping 1 t. of hot syrup into separate clear cup of ice water.

3. 248 °F to 252 °F – Follow basic procedure for the cold water test, but continue to heat syrup to 248 °F to 252 °F (120-122 °C) before dropping 1 t. of hot syrup into separate clear cup of ice water.

4. 260 °F to 270 °F – Follow basic procedure for the cold water test, but continue to heat syrup to 260 °F to 270 °F (127-132 °C) before dropping 1 t. of hot syrup into separate clear cup of ice water.

5. 300 °F to 310 °F – Follow basic procedure for the cold water test, but continue to heat syrup to 300 °F to 310 °F (149-154 °C) before dropping 1 t. of hot syrup into separate clear cup of ice water.

Table A-1 Cold Water Test for Sugar Syrup			
Final temperature	**Appearance of syrup**	**Color of syrup**	**Texture of syrup**
234 °F to 240 °F			
244 °F to 248 °F			
248 °F to 252 °F			
260 °F to 270 °F			
300 °F to 310 °F			

Table A-2 – Cold Water Test for Sugar Syrups		
Final temperature of syrup	**Description of test**	**Product**
234 °F to 240 °F	**Soft ball** – Syrup forms a soft ball that flattens on removal from water	Fondant Fudge
244 °F to 248 °F	**Firm ball** – Syrup forms a firm ball that does not flatten on removal from water	Caramels
248 °F to 252 °F	**Hard ball** – Syrup forms a ball that is hard enough to hold its shape, yet plastic	Divinity Marshmallows
260 °F to 270 °F	**Soft crack** – Syrup separates into threads that are hard but not brittle	Butterscotch Taffy
300 °F to 310 °F	**Hard crack** – Syrup separates into threads that are hard and brittle	Brittle Toffee

B. *Preparation of Fondant*

Objectives

1. To gain experience in the preparation of a crystalline candy.
2. To become familiar with the cold water test.
3. To determine the influence of corn syrup on the texture of the finished fondant.

Basic Fondant Recipe

Ingredients

1 c. sugar
1 T. corn syrup
¼ c. + 1 T. water
1 c. ice water in clear glass (for cold water test)

Procedures

1. Place ¼ c. + 1 T. water, sugar, and corn syrup in a heavy small saucepan over low heat. Position a candy thermometer so the bulb is immersed in the mixture.
2. Stir gently until sugar is dissolved. Avoid splashing on sides of saucepan above level of mixture.
3. Cover; bring to a boil, cooking gently for about 3 minutes. Do not stir.
4. Remove cover; continue cooking gently to 238 °F (114 °C) on a candy thermometer. Perform cold water test.
5. Remove from heat; pour onto a large, wet platter. Do not scrape syrup from the pan. Do not agitate until the temperature drops to 120 °F (49 °C).
6. Once temperature drops to 120 °F, stir until syrup turns creamy white. Knead the mixture to produce the desired consistency. Fondant should be smooth and creamy white. Transfer fondant to a plate.
7. Perform sensory evaluation and record results in Table B-1.

Variations

1. Control – Prepare basic recipe for fondant as above.

2. No interfering agent – Follow basic recipe for fondant, except omit the corn syrup.

Table B-1 Preparation of Fondant

Candy	Appearance	Flavor	Texture
Fondant with corn syrup			
Fondant without corn syrup			

C. Preparation of Fudge

Objectives
1. To gain experience in the preparation of a crystalline candy.
2. To become familiar with the cold water test.
3. To observe the effect of syrup temperature at the time of beating on crystallization in fudge.

Basic Recipe for Fudge

Ingredients

1 c. granulated sugar
½ c. whole milk
½ T. light corn syrup
1 oz. unsweetened chocolate, cut into pieces
1 T. butter
½ t. vanilla
1 c. ice water in clear glass (for cold water test)

Procedure
1. Lightly butter a small pie plate or baking dish.
2. Blend together the sugar, corn syrup, and milk in a small saucepan. Position a candy thermometer so that the bulb is immersed in the mixture.
3. Heat the mixture over medium heat until it boils. Stir constantly to prevent scorching.
4. Reduce heat slightly and continue cooking, stirring slowly, until 234 °F (112 °C), the soft-ball stage, is reached. Perform the cold water test. Be sure no sugar crystals are clinging to the side of the saucepan.
5. Remove the pan from the heat and add the butter and chocolate. Do not stir the mixture. Allow the cooked syrup to cool to 120 °F (49 °C).
6. Remove the candy thermometer and add the vanilla.
7. Using a wooden spoon, beat the mixture until the fudge loses its gloss and appears ready to set.
8. Quickly press the mixture into the buttered dish.
9. Evaluate the appearance, flavor and texture of the fudge and record observations in Table C-1.

Variations

1. Control – Prepare basic recipe for fudge as above.

2. Modified temperature – Prepare basic recipe for fudge, except cool the cooked fudge mixture to 176 °F (80 °C) before beginning to beat.

Table C-1 Preparation of Fudge

Beating temperature	Appearance	Flavor	Texture
Fudge – 120 °F			
Fudge – 176 °F			

D. Using Sugar Substitutes in Preparation of Noncrystalline Candy

Objectives
1. To gain experience in the preparation of noncrystalline candy.
2. To emphasize the importance of granulated sugar in the successful preparation of noncrystalline candy.
3. To gain experience preparing noncrystalline candy with a sugar substitute.

Basic Recipe for Noncrystalline Candy

Ingredients

1 c. granulated sugar
¼ c. 2 T. corn syrup
½ c. water
½ t. lemon or mint extract
Food coloring (optional)
Toothpicks (optional)
1 c. ice water in clear glass (for cold water test)
Oil (for oiling surface)

Procedure
1. Combine the sugar, corn syrup, and water in a saucepan. Position a candy thermometer so that the bulb is immersed in the mixture.
2. Heat the mixture over medium-high heat. Stir constantly and continue heating until the mixture boils.
3. Reduce heat to medium-low and continue cooking to 310 °F (154° C), the hard-crack stage. Lower the heat near the end of the cooking period. Check doneness with the cold water test.
4. When the syrup has reached the endpoint temperature, remove from heat and immediately stir in the flavoring and coloring.
5. With a spoon, drop the hot syrup onto an oiled cookie sheet in bite-sized pieces. If desired, lollipops can be made by placing toothpicks into the candy before it hardens.
6. Before the syrup completely hardens, loosen candy from the oiled surface with a spatula.
7. Evaluate the appearance, flavor, and texture of the candy and record in Table D-1.

Variations

1. Control, granulated sugar – Prepare basic recipe for noncrystalline candy as above.

2. Sugar substitute – Prepare basic recipe for noncrystalline candy, except replace sugar with equivalent amount of sugar substitute (1 c. Splenda®).

Table D-1 Using Sugar Substitute in Preparation of Noncrystalline Candy

Candy	Appearance	Flavor	Texture
Noncrystalline candy – granulated sugar			
Noncrystalline candy – sugar substitute			

E. Evaluation of Nonnutritive Sweeteners

Objectives

1. To become familiar with various nonnutritive sweeteners.
2. To compare the appearance, flavor, and texture of nonnutritive sweeteners.
3. To compare nonnutritive sweeteners to granulated sugar.

Basic Procedure to Evaluate Nonnutritive Sweeteners

Ingredients

Assortment of nonnutritive sweeteners (chosen by instructor)
Fruit (chosen by instructor)

Procedure

1. Place selected nonnutritive sweeteners in individual bowls.
2. Cut bite-size pieces of fruit and place on a tray.
3. Sprinkle a small amount of nonnutritive sweetener on pieces of fruit for sampling.
4. Evaluate each nonnutritive sweetener for appearance, flavor, and texture. Note ingredients, kcalories per serving, and cost of nonnutritive sweeteners. Record all observations in Table E-1.

Table E-1 Evaluation of Nonnutritive Sweeteners

Product	Appearance	Flavor	Texture	kcal/ serving	Ingredients	Price/ oz.

Additional Food Preparation Exercises

F. Preparation of Noncrystalline Candies

Objectives

1. To gain experience in the preparation of noncrystalline candy.

F-a. Peanut Brittle

Peanut Brittle

Ingredients

1 ¼ c. sugar
½ c. corn syrup
½ c. water
2 T. butter or margarine
½ t. baking soda
1 c. unsalted peanuts
Ice water (for cold water test)

Procedure

1. Combine sugar, water, and corn syrup in a saucepan. Position a candy thermometer so that the bulb is immersed in the mixture.
2. With slow stirring, heat the mixture rapidly over medium-high heat to 280 °F. Add the peanuts and the butter. Stir the mixture continuously and continue heating to 306 °F, the hard-crack stage. Check doneness with the cold water test. Remove pan from heat immediately.
3. Quickly stir in the baking soda. Do not over-stir. Be careful not to let the mixture overflow.
4. Pour onto a greased cookie sheet, forming a large square of brittle. Cool and then break into pieces.

F-b. Toffee

Toffee

Ingredients

1 c. sugar
¾ c. 2 T. butter
¼ c. water
2 T. corn syrup
¼ c. toasted, sliced almonds (optional)
Ice water (for cold water test)

Procedure

1. Combine all ingredients except almonds in a saucepan. Position a candy thermometer so that the bulb is immersed in the mixture.
2. Stirring slowly, heat the mixture over medium-high heat until the sugar is dissolved and the mixture comes to a boil.
3. Stirring slowly to prevent scorching, reduce heat and continue boiling to 300 °F, the hard-crack stage. Check doneness with the cold water test.
4. Remove from heat and stir in the almonds.
5. Pour the candy into a buttered baking dish or pie plate.
6. When cool, break into pieces.

G. Pralines

Pralines

Ingredients
1 pkg. brown sugar
2 sticks margarine, melted
2 eggs
1 ½ c. flour
1 t. vanilla extract
1 c. pecans
Aerosol vegetable oil cooking spray

Procedure
1. Preheat oven to 350 ºF. Spray pan with aerosol vegetable oil cooking spray and set aside.
2. Place brown sugar in a large mixing bowl.
3. Melt margarine and pour over brown sugar. Beat margarine and brown sugar together well.
4. Add eggs to brown sugar mixture and mix well. Mix in flour and vanilla extract.
5. Fold nuts into mixture.
6. Place mixture in prepared pan and bake for 30 minutes.

From the recipe file of Alma Mynhier

H. Chocolate Covered Cherries

Chocolate Covered Cherries

Ingredients
2 lbs. powered sugar
14 oz. condensed milk
½ c. margarine or butter
~3 doz. maraschino cherries
12 oz. chocolate chips
2 T. paraffin, grated

Procedure
1. Soften margarine at room temperature.
2. Drain maraschino cherries in a colander.
3. Mix powdered sugar, condensed milk, and margarine together.
4. Portion approximately 1 T. powdered sugar mixture and flatten in gloved hand. Place a drained cherry in the center of the mixture. Wrap the mixture around the cherry.
5. Place the wrapped cherry on a wax paper-lined pan. Insert a toothpick in the center of the cherry. Repeat this process for all cherries.
6. Cover the pan of cherries with foil and chill in the refrigerator for a minimum of two hours.
7. Melt chocolate chips and paraffin in the top of a double boiler.
8. Dip cherries in melted chocolate mixture and place on wax paper to harden. Remove toothpick, place small amount of chocolate over toothpick hole, and allow to harden.

From the recipe file of Karen Beathard

Post-Lab Study Questions

Textbook reference: Chapters 21 and 25

Discussion Questions

1. Explain the mechanism by which cream of tartar acts as an interfering agent.

2. If water boils at 209 °F on your candy thermometer, to what temperature should you cook your fudge mixture?

3. Is a candy thermometer or the cold water test a more reliable index to doneness? Explain.

4. Why should the beating of a supersaturated fondant or fudge syrup be uninterrupted?

5. Why do you think fudge and fondant become more firm if they are chilled in the refrigerator?

Questions for Post-Lab Writing

6. Compare the fondant with and without corn syrup.

7. Compare the fudge beaten at 120 °F and 176 °F.

8. Explain the mechanism of the interfering agents in each of the following: fondant, fudge, hard candies, peanut brittle, and toffee.

9. What is the source of the acid that reacts with the baking soda in peanut brittle?

10. Why do you think many crystalline candies become sticky when exposed to the air?

11. Why do you think flavoring agents are added after the candy syrup has finished cooking?

Unit 19 – Frozen Desserts

Introduction

The preparation of frozen desserts involves the controlled formation of ice crystals to achieve a desirable texture. Small crystals contribute to a smooth texture. Agitating the frozen dessert mix during crystallization aids in the formation of many small crystals. The ingredients in the mix and the amount of agitation are both factors that influence the texture of frozen desserts.

The freezing point or melting point of a substance is the temperature at which a liquid and its solid form exist in equilibrium. When a solute is added to water, the freezing point of the solution is lower than that of pure water. The application of this principle is an important part of frozen dessert preparation.

Basic agitated frozen dessert ingredients include liquid (milk, cream, juice) and sugar. Optional ingredients include egg yolk, stabilizers, flavorings, and emulsifiers. The frozen dessert mix is placed in a metal container, which is then surrounded by a mixture of rock salt and crushed ice. The ratio of salt and ice used depends on the specific frozen dessert. The salt : ice ratio for ice creams is 1:8 by weight (¼ c. rock salt to 1 qt. ice). For sherbets, the salt : ice ratio is 1:6 by weight ($^1/_3$ c. rock salt to 1 qt. ice) because sherbets have a lower freezing point than ice creams and therefore require a colder brine bath. Heat from the frozen dessert mix is conducted through the metal container and melts the ice in the brine bath. As the ice melts, salt dissolves in the water formed. This process continues until all the salt is in solution. The equilibrium temperature of the salt-water solution is 10 °F to 14 °F. Since the heat used to melt the ice comes from the frozen dessert mix, the end result is a firm frozen dessert mix in which approximately three-fourths of the water has been converted into ice crystals. The frozen dessert mix is constantly agitated during the cooling process so that many small ice crystals are formed, rather than a few ice crystals, resulting in a smooth texture.

Interfering agents, such as high-fat ingredients, beaten egg whites, gelatin, corn syrup, and whipped cream, are often included in greater quantities in the preparation of still-frozen (non-agitated) desserts to prevent the formation of large crystals. A higher proportion of interfering agents is needed in still-frozen desserts because the ice crystals are not kept small by agitation.

Overrun, the volume of frozen dessert obtained in excess of the volume of the mix, is due to both the expansion of water as it crystallizes and the air incorporated into the mix. Overrun is of interest because of its relationship to yield and because of its influence on mouthfeel and texture. Too much overrun causes the ice cream to be frothy. Approximate overrun is 35-50% for homemade ice creams, 70-100% for commercial ice creams, and 30-40% for sherbets.

Key Terms

body – describes the consistency of frozen desserts, considering firmness, richness, viscosity, and resistance to melting.

dasher – equipment placed within an ice cream freezer that stirs the ice cream mix while it freezes. The speed of the dasher determines how much air is incorporated into the mixture.

overrun – increase in volume due to incorporation of air into a frozen dessert, resulting in a softer and creamier product.

sorbet – frozen dessert prepared with pureed fruit or fruit flavoring and sugar syrup. Sorbet does not contain fat, eggs, gelatin, or dairy products.

still-frozen dessert – light, airy, smooth dessert that is not agitated during freezing and includes whipped egg whites or whipped cream, which prevents large ice crystal formation.

Pre-Lab Questions

Textbook reference: Chapter 26

1. Can crushed ice alone freeze ice cream? Why is rock salt used for homemade ice cream preparation?

2. How does agitation affect the texture of frozen desserts?

3. What interfering agents are used in ice cream preparation?

4. How does the percent overrun in commercial products compare to that of homemade ice cream?

5. Explain why heat shock is a problem for ice cream manufacturers.

Lab Procedures

A. Vanilla Ice Cream

Objectives

1. To gain experience in the preparation of an agitated frozen dessert.
2. To interpret the effect of varying levels and types of sweeteners on the freezing point of ice cream mixtures.
3. To define and calculate overrun in ice creams and identify the factors that contribute to overrun.
4. To illustrate the effect of a stabilizer on the body of ice cream.
5. To illustrate the effect of a stabilizer on the texture of ice cream during storage.

Basic Recipe for Vanilla Ice Cream

Ingredients

½ c. granulated sugar
1 ½ c. whole milk
1 c. whipping cream
1 ½ t. vanilla
Pinch salt
Crushed ice (for ice cream maker)
Rock salt (for ice cream maker)

Procedure

1. Combine sugar, salt, and milk in a saucepan. While stirring continuously, heat mixture over medium heat until the sugar is dissolved.
2. Remove from heat and add vanilla and cream. Stir well.
3. Measure the volume of the mix.
4. Pour mixture into chilled freezer can from an electric or hand-cranked ice cream maker, filling no more than $^2/_3$ full.
5. Freeze ice cream according to ice cream freezer manufacturer directions. Record time required to freeze ice cream.
6. Remove the lid and dasher. Scrape the ice cream on the dasher back into the can.
7. Using a glass measuring cup, measure the volume of the ice cream. Then quickly transfer the ice cream back to the freezer can.
8. Cover ice cream and store in freezer for approximately 30 minutes, if time permits.
9. Calculate the **percent overrun** using the following equation:

$$\% \text{ overrun} = \frac{\text{volume (frozen)} - \text{volume (unfrozen)} \times 100}{\text{volume (unfrozen)}}$$

10. Evaluate appearance, flavor, and texture and record observations in Table A-1.
11. Transfer some of the ice cream into a storage container, store in the freezer for 1 week, and then reevaluate its appearance, flavor, and texture.

Variations

1. Control – Prepare basic recipe for vanilla ice cream as above.

2. Doubled granulated sugar – Follow the basic recipe for vanilla ice cream, except double the amount of granulated sugar.

3. Sugar substitute – Follow the basic recipe for vanilla ice cream, except replace granulated sugar with ½ c. sugar substitute.

4. Honey – Follow the basic recipe for vanilla ice cream, except replace granulated sugar with ½ c. honey.

5. Plain unflavored gelatin – Follow the basic recipe for vanilla ice cream, except blend 1 envelope (7 g) of plain unflavored gelatin with the granulated sugar.

Table A-1 Homemade Vanilla Ice Cream

Ice cream	Freezing time	% overrun	Flavor	Texture (mouthfeel)	Body (melting behavior)
Control, basic recipe					
1 c. sugar					
Sugar substitute					
Honey					
Basic recipe with gelatin					

B. *Comparing Sherbet and Low-Fat Ice Cream*

Objectives

1. To gain experience in the preparation of agitated frozen desserts.
2. To become familiar with sherbet and low-fat ice cream and to compare them with ice cream.

B-a. Orange Sherbet

Basic Recipe for Orange Sherbet

Ingredients

2 c. whole milk
½ c. frozen orange juice concentrate
½ c. sugar
½ t. finely grated orange rind
Pinch of salt
Crushed ice (for ice cream maker)
Rock salt (for ice cream maker)

Procedure

1. Combine ½ c. milk, sugar, and salt in a saucepan. While stirring continuously, heat mixture over medium heat until the sugar is dissolved.
2. Remove from heat and cool at least to room temperature.
3. Add remaining ingredients. Stir well.
4. Measure the volume of the mix.
5. Pour mixture into chilled freezer can from an electric or hand-cranked ice cream maker, filling no more than $^{2}/_{3}$ full.
6. Freeze sherbet according to ice cream freezer manufacturer directions. Record time required to freeze sherbet.
7. Remove the lid and dasher. Scrape the sherbet on the dasher back into the can.
8. Using a glass measuring cup, measure the volume of the sherbet. Then quickly transfer the sherbet back to the freezer can.
9. Cover sherbet and store in freezer for approximately 30 minutes, if time permits.
10. Calculate the **percent overrun** using the following equation:

$$\% \text{ overrun} = \frac{\text{volume (frozen)} - \text{volume (unfrozen)} \times 100}{\text{volume (unfrozen)}}$$

11. Evaluate appearance, flavor, and texture and record observations in Table B-1.

B-b. Low-Fat Vanilla Ice Cream

Basic Recipe for Low-Fat Vanilla Ice Cream

Ingredients

2 ½ c. 2% milk
½ c. granulated sugar
1 ½ t. vanilla
Pinch salt
Crushed ice (for ice cream maker)
Rock salt (for ice cream maker)

Procedure

1. Combine ½ c. milk, sugar, and salt in a saucepan. While stirring continuously, heat mixture over medium heat until the sugar is dissolved.
2. Remove pan from heat. Add remaining milk and vanilla. Stir well.
3. Measure the volume of the mix.
4. Pour mixture into chilled freezer can, filling no more than $^{2}/_{3}$ full.
5. Freeze low-fat vanilla ice cream according to ice cream freezer manufacturer directions. Record time required to freeze the ice cream.
6. Remove the lid and dasher. Scrape the ice cream on the dasher back into the can.
7. Using a glass measuring cup, measure the volume of the ice cream. Then quickly transfer the ice cream back to the freezer can.
8. Cover ice cream and store in freezer for approximately 30 minutes, if time permits.
9. Calculate the **percent overrun** using the following equation:

$$\% \text{ overrun} = \frac{\text{volume (frozen)} - \text{volume (unfrozen)} \times 100}{\text{volume (unfrozen)}}$$

10. Evaluate appearance, flavor, and texture and record observations in Table B-1.

Table B-1 Comparing Sherbet and Low-Fat Ice Cream

Product	Freezing time	% overrun	Flavor	Texture (mouthfeel)	Body (melting behavior)
Sherbet					
Low-fat ice cream					

Additional Food Preparation Exercises

C. Non-Agitated Frozen Desserts

Objectives
1. To prepare several types of non-agitated frozen desserts.
2. To emphasize the use of interfering agents in the preparation of non-agitated frozen desserts.

C-a. Frozen Yogurt

Frozen Yogurt

Ingredients
½ envelope unflavored gelatin
½ T. lemon juice
2 T. cold water
1 c. strawberries, peaches, raspberries, or cantaloupe
1 T. granulated sugar
1 c. vanilla yogurt

Procedure
1. Sprinkle the gelatin over 2 T. cold water in the top of a double boiler.
2. Heat this mixture over hot water, stirring just until the gelatin dissolves.
3. Remove mixture from heat.
4. Add the fruit, yogurt, sugar, and lemon juice and mix lightly.
5. Pour the mixture into a blender and blend until it is fairly smooth. The fruit may remain in small pieces for color and texture.
6. Pour the mixture into a shallow metal pan. Cover the pan with foil and freeze until frozen.
7. Evaluate during the next laboratory period.

C-b. Fruit Bavarian Cream

Fruit Bavarian Cream

Ingredients
1 envelope unflavored gelatin
¼ c. cold water
¼ c. boiling water or juice from canned fruit
¾ c. sugar
¾ c. crushed fruit
¾ c. whipping cream

Procedure
1. Bring water or fruit juice to boil in a saucepan and remove from heat.
2. Sprinkle the gelatin over the cold water in a small bowl. Disperse gelatin mixture in the boiling water or fruit juice.
3. Add the sugar and stir until dissolved. Transfer mixture to mixing bowl.
4. Stir in the fruit. Chill the mixture until it begins to thicken.
5. Whip the cream in a separate mixing bowl until stiff peaks form. Fold whipping cream into the fruit mixture.
6. Pour into a gelatin mold or metal pan. Chill until firm.
7. Turn out of the mold. Garnish with pieces of fruit, if desired.

Post-Lab Study Questions

Textbook reference: Chapter 26

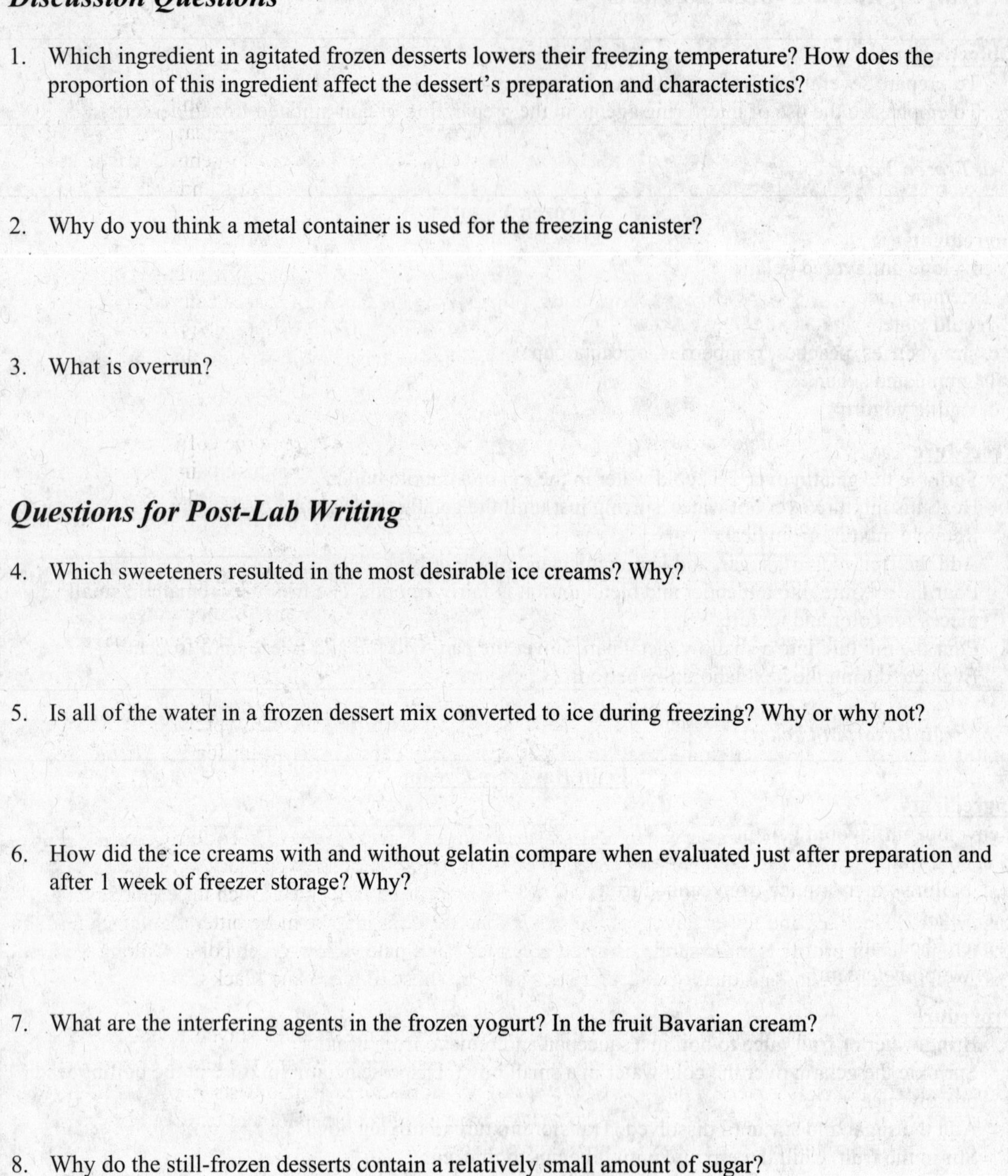

Discussion Questions

1. Which ingredient in agitated frozen desserts lowers their freezing temperature? How does the proportion of this ingredient affect the dessert's preparation and characteristics?

2. Why do you think a metal container is used for the freezing canister?

3. What is overrun?

Questions for Post-Lab Writing

4. Which sweeteners resulted in the most desirable ice creams? Why?

5. Is all of the water in a frozen dessert mix converted to ice during freezing? Why or why not?

6. How did the ice creams with and without gelatin compare when evaluated just after preparation and after 1 week of freezer storage? Why?

7. What are the interfering agents in the frozen yogurt? In the fruit Bavarian cream?

8. Why do the still-frozen desserts contain a relatively small amount of sugar?

9. Why are the still-frozen desserts put into a shallow metal pan during freezing?

Unit 20 – Beverages

Introduction

There are two objectives in the preparation of a good cup of coffee or tea. One is to optimize the extraction of desirable, water-soluble flavor constituents from the coffee bean or tea leaf and retain them in the beverage. The other is to minimize the extraction of undesirable favor constituents. To achieve these objectives, grind of coffee, size of tea leaves, contact time, water temperature, and brewing methods must be controlled.

The coffee grind determines the proper proportions of coffee and water. When a fine grind of coffee is used, more surface area is available for contact with water. The greater surface area can result in more efficient extraction of flavor compounds such as tannins, volatile oils, and organic acids. Therefore, a smaller proportion of coffee to water is required to yield brews of acceptable strength than if a large grind is used.

Because the flavor compounds are extracted by water, the contact time between the coffee or tea and water will influence the final beverage characteristics. Too long a contact time results in a bitter beverage due to the extraction of large amounts of tannins. Too brief a contact time results in a flat, flavorless beverage since desirable oils and organic acids are incompletely extracted.

The optimum brewing temperature for both coffee and tea is 185-203 °F. A brewing temperature greatly above 203 °F results in bitter coffee because more tannins are extracted at higher temperatures and volatile flavor compounds are lost. A lower brewing temperature produces a flat, flavorless beverage since organic acids and oils are not maximally extracted.

Coffee is commonly brewed by filtering (see Figure 20-1) and is only occasionally prepared by steeping in the United States. Tea is commonly prepared by steeping tea leaves in hot water for 3-5 minutes (see Figure 20-2).

Black, green, and oolong teas are available. Black tea refers to leaves that have been fermented, resulting in a decrease in tannins and an increase in desirable flavor compounds. Brewed black tea has an amber color and a full flavor. Green tea refers to leaves that have not been fermented. Such tea contains more tannins than black tea and fewer flavor compounds. Some varieties may be more bitter than black tea, and the overall flavor profile is more subtle. Brewed green tea has a pale yellow-green color. Oolong tea is a partly fermented tea and has quality characteristics between those of green and black teas.

Black tea leaves are classified by their size. Orange pekoe refers to the smallest leaves, pekoe to the middle-size leaves, and souchong to the largest leaves. "Broken" simply means the leaves are not intact. Broken orange pekoe is generally considered the best grade of black tea and consists chiefly of buds. The buds and the first two leaves of the harvested tea shoot have a high polyphenol and enzyme content, which contribute to the color and flavor of the fermented tea.

Chocolate and cocoa are produced by fermenting, roasting, cracking, and grinding the seeds of the pods of the cacao tree. The resulting paste is called chocolate liquor and solidifies on cooling to form bitter chocolate. Milk, sugar, and flavorings are added to produce milk chocolate. Chocolate contains approximately 50% fat (cocoa butter). In the production of cocoa, the cocoa butter is pressed from the chocolate liquor. The solid material remaining is dried and ground to a fine cocoa powder. Cocoa and chocolate are classified as either natural-processed or Dutch-processed. Dutch-processed means that the seeds have been treated with alkali to improve color and solubility. Dutch-processed products have a slightly milder, less bitter flavor and darker color than natural-processed products.

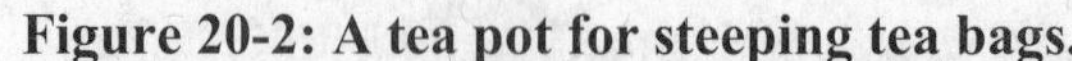

Figure 20-1: A drip coffee pot.

Figure 20-2: A tea pot for steeping tea bags.

Quality Characteristics of Coffee, Tea, Chocolate, and Cocoa Beverages			
	Coffee	**Tea**	**Chocolate and Cocoa**
Appearance	• clear, bright • deep amber to rich brown	• clear, bright • black tea: amber • green tea: yellow-green • depth of color is dependent on time of steeping	• surface free from milk "skin" or fat globules
Aroma	• highly fragrant • pleasing aroma	• distinct but subtle	• definite chocolate aroma
Flavor	• typical coffee flavor with no bitterness or staleness • mellow taste	• characteristic of type • green tea may be more bitter than black tea	• clean, characteristic chocolate flavor
Mouthfeel	• smooth feel on tongue • slightly astringent	• astringent • dependent on type and length of brewing	• smooth consistency • body similar to thin cream

Key Terms

astringent – sour or bitter in taste, resulting in the puckering of mouth tissues.

mellow – pleasing and full of flavor due to aging.

steep – to let stand in hot liquid below boiling temperature to extract flavor, color, or other qualities from a specific food.

strain – to separate particles from liquid through use of a screen, sieve, colander, etc.

Pre-Lab Questions

Textbook reference: Chapter 27

1. Name some factors that may influence the strength of brewed coffee.

2. What contribution does caffeine make to the flavor of coffee?

3. Why would you expect flavor differences between brewed and instant or freeze-dried beverages?

4. What are the optimum temperature and duration for steeping tea?

5. Which components of black tea contribute to the flavor of the beverage? Which contribute to the color of the beverage?

Lab Procedures

A. Coffee Brewing Methods

Objectives

1. To observe different methods of brewing coffee.
2. To compare the quality of coffee beverages prepared by different brewing methods.
3. To compare the amounts of tannins present in coffee brewed by different methods.

Basic Recipe for Coffee Preparation

Ingredients

2 T. coffee
1 c. water

Procedure

1. Use a drip (filter) coffee maker. Put a filter into the filter holder and add drip-grind coffee.
2. Pour water into the water reservoir. Turn coffee pot on.
3. Record the temperature of the water in the filter holder. Remove the filter and the coffee grounds when the coffee has finished dripping into the pot.
4. Record the temperature of the brewed coffee.
5. Evaluate the color, aroma, and flavor. Record all data and observations in Table A-1.

Variations

1. Drip (filter) coffee – Follow basic recipe for coffee preparation as above.

2. Steeped coffee – Follow basic recipe for coffee preparation, except do not use a standard coffee pot. Boil water in a saucepan. Remove saucepan from heat, add individual packet of coffee, and stir well. Steep 3-5 minutes. Record the temperature of the steeped coffee as steeping proceeds. If the temperature drops below 185 °F, heat gently but do not exceed 203 °F. Carefully strain coffee through cheesecloth. Record observations in Table A-1.

3. Decaffeinated drip (filter) coffee – Follow the basic recipe for coffee preparation, except substitute decaffeinated drip-grind coffee for caffeinated drip-grind coffee.

Table A-1 Coffee Brewing Methods

Method	Color	Aroma	Flavor	Temperature	
				Start	End
Drip (filter)					
Steeped					
Decaffeinated drip (filter)					

B. *Comparison of Instant Coffees*

Objectives

1. To compare regular, decaffeinated and freeze-dried instant coffees.
2. To compare brewed coffee (from part A) with instant coffee.

Basic Recipe for Instant Coffee

Ingredients

1 serving regular instant coffee
Boiling water

Procedure

1. Prepare 1 c. of medium-strength regular instant coffee according to the directions on the manufacturer's label.
2. Evaluate the color, clarity, aroma, and flavor of the coffee. Record observations in Table B-1.

Variations

1. Medium-strength regular instant coffee – Follow the basic recipe for instant coffee as above.

2. Freeze-dried coffee – Follow the basic recipe for instant coffee, except substitute freeze-dried coffee for regular instant coffee.

3. Decaffeinated instant coffee – Follow the basic recipe for instant coffee, except substitute decaffeinated instant coffee for regular instant coffee.

4. Decaffeinated freeze-dried coffee – Follow the basic recipe for instant coffee, except substitute decaffeinated freeze-dried coffee for regular instant coffee.

Table B-1 Comparison of Instant Coffees

Coffee	Color	Clarity	Aroma	Flavor
Regular instant				
Regular freeze-dried				
Decaffeinated instant				
Decaffeinated freeze-dried				

C. Comparison of Teas

Objectives
1. To gain experience in brewing tea.
2. To compare black, oolong, and green teas for quality characteristics.

Basic Recipe for Tea Preparation

Ingredients

1 t. green tea leaves
1 c. boiling water

Procedure
1. Preheat a teapot by filling it with boiling water, allowing it to stand for 3 minutes, and then pouring out the water.
2. Place tea leaves in the preheated teapot, and pour boiling water over them.
3. Steep 3 minutes. Record the temperature at the end of steeping. Then strain the tea into a cup.
4. Evaluate color, aroma, astringency, and flavor. Record all observations and data in Table C-1.

Variations

1. Green tea leaves – Follow basic recipe for tea preparation above.

2. Oolong tea leaves – Follow basic recipe for tea preparation, except use oolong tea leaves.

3. Black tea leaves – Follow basic recipe for tea preparation, except use black tea leaves.

4. Decaffeinated tea leaves – Follow basic recipe for tea preparation, except use decaffeinated tea leaves.

5. Herbal tea leaves – Follow basic recipe for tea preparation, except use herbal tea leaves.

Table C-1 Comparison of Teas

Tea	Color	Aroma	Astringency	Flavor
Green				
Oolong				
Black				
Decaffeinated				
Herbal				

D. Effect of Acid on Tea

Objectives

1. To compare the color and clarity of brewed teas with and without the addition of lemon juice.

Basic Procedure to Evaluate Effect of Acid on Tea

Ingredients

2 black tea bags
2 c. boiling water
1 T. lemon juice

Procedure

1. Steep 2 black tea bags in 2 cups of boiling water for 4 minutes. Remove tea bags and divide the tea equally among 4 custard cups.
2. Add ½ T. lemon juice to 2 custard cups.
3. Stir each cup well.
4. Refrigerate 1 of the cups of tea without lemon juice and 1 of the cups of tea with lemon juice for 50 minutes each.
5. Compare the room-temperature and chilled teas with and without lemon juice for clarity and color. Record observations in Table D-1.

Table D-1 Effect of Acid on Tea

Tea Variation	Clarity	Color
Room temperature, plain		
Room temperature, with lemon juice		
Chilled, plain		
Chilled, with lemon juice		

E. Comparison of Chocolate and Cocoa Beverages

Objectives

1. To compare homemade chocolate and cocoa beverages.
2. To compare commercial and homemade cocoa beverages.
3. To observe the effects of heat on cocoa and chocolate.

Basic Recipe for Cocoa

Ingredients

2 T. cocoa
2 T. granulated sugar
½ c. water
1 ½ c. milk
¼ t. vanilla
Dash of salt

Procedure

1. Mix together cocoa, sugar, and salt in a small saucepan.
2. Add water and blend well.
3. Heat to boiling, stirring constantly.
4. Continue heating and stirring until mixture thickens and is smooth. Be careful not to allow the mixture to scorch.
5. Add milk and heat to 201 °F.
6. Add vanilla and stir well.
7. Evaluate product and record results in Table E-1.

Variations

1. Homemade cocoa beverage – Follow basic recipe for cocoa as above.

2. Homemade chocolate beverage – Follow basic recipe for cocoa, except substitute ½ square unsweetened chocolate for cocoa.

3. Commercial cocoa beverage – Prepare 2 packages commercial instant cocoa beverage according to package directions.

Table E-1 Comparison of Chocolate and Cocoa Beverages

Beverage	Color	Aroma	Flavor	Consistency
Homemade cocoa				
Homemade chocolate				
Commercial cocoa				

Additional Food Preparation Exercises

F. Preparation of Selected Beverages

Objectives
1. To gain experience in beverage preparation.
2. To become familiar with a variety of beverages.

F-a. Viennese Coffee Mix

Viennese Coffee Mix

Ingredients
¼ c. hot cocoa mix
¼ c. nondairy creamer mix
3 T. instant coffee powder
2 T. powdered sugar
$^{1}/_{8}$ t. cinnamon
$^{1}/_{8}$ t. nutmeg

Procedure
1. Combine all ingredients.
2. Add 3 T. to a cup. Fill with boiling water, stir, and serve.

F-b. Cafe Aruba

Café Aruba

Ingredients
1 orange
½ c. coffee
1 T. sugar (divided)
1 qt. water
½ c. heavy cream

Procedure
1. With a vegetable peeler, remove the colored peel from the orange in thin strips. Make the strips very thin, avoiding the bitter white underneath. Set aside $^{1}/_{3}$ of the peelings for garnish.
2. Cut the remaining $^{2}/_{3}$ of the peelings into large pieces.
3. Combine the large pieces of orange peelings, the coffee, and 1 ½ t. sugar.
4. Prepare coffee using the indicated amount of water.
5. Meanwhile, whip cream until it is almost thickened. Add in remaining 1 ½ t. sugar. Continue beating until it has the appearance of whipped cream.
6. Finish peeling orange and cut half of the orange into thin crosswise slices. (The other half of the orange is not needed.)
7. When coffee is brewed, remove coffee basket and add orange slices. Let it mellow 10 minutes before serving.
8. Remove orange slices and pour coffee into serving cup. Cover with approximately ¼ of the whipped cream and garnish with an orange peel.

F-c. Hot Spiced Tea

Hot Spiced Tea

Ingredients

1 qt. boiling water
¼ c. sugar
10 whole cloves
2 cinnamon sticks
4 tea bags
¼ c. orange juice
1 T. orange rind, grated
½ lemon, sliced thin and seeded

Procedure

1. Combine the water, sugar, cloves, and cinnamon sticks in a saucepan.
2. Boil the mixture for 1 minute.
3. Remove from heat, add the tea bags, and steep for 4 minutes, covered.
4. Remove tea bags, add last 3 ingredients, and place over low heat to keep warm. Hold below the boiling point for about 5 minutes.
5. Strain and serve in beverage cups.

F-d. French Chocolate

French Chocolate

Ingredients

1 ¼ square bitter chocolate
3 T. water
¼ c. + 2 T. sugar
½ c. heavy cream, whipped
1 qt. milk, scalded

Procedure

1. Cook the chocolate with water until thick, stirring as necessary to prevent sticking or burning.
2. Add the sugar and let the mixture boil up once. Remove from heat and cool.
3. Fold whipped cream into cooled chocolate mixture.
4. Place a heaping tablespoon of the chocolate-whipped-cream mixture in the bottom of a beverage cup. Fill the cup with hot milk. Serve at once.

F-e. Mexican Chocolate

Mexican Chocolate

Ingredients

2 c. milk
1 cinnamon stick
1 ½ square semisweet chocolate

Procedure

1. Heat the milk and cinnamon slowly over low heat, stirring occasionally to prevent sticking.
2. When the milk is scalding hot, add the chocolate.
3. When the chocolate has melted, remove the mixture from the heat and beat vigorously with an electric mixer.
4. Serve in beverage cups, making sure that each cup is topped with foam.

F-f. Orange Julius

Orange Julius

Ingredients

6 oz. frozen orange juice concentrate
½ c. sugar
1 c. water
1 c. milk
1 t. vanilla
Ice

Procedure

1. Mix all ingredients in a blender.
2. Add ice to the 6 c. mark on the blender.
3. Blend all ingredients together. Serve at once. Yield: 7 c.

From the recipe file of Lesha Emerson

Post-Lab Study Questions

Textbook reference: Chapters 25 and 27

Discussion Questions

1. What is the effect on brewed coffee if the temperature of the water is above or below 185-203 °F?

2. Compare the contact time between water and grounds for the brewing methods used.

3. Why are coffee makers of glass or porcelain usually recommended?

4. Was the temperature of the water in the preheated pots within the range recommended for brewing tea after 3 minutes of steeping?

5. Compare the beverages made from green tea, black tea, oolong tea, decaffeinated tea, and herbal tea.

6. How is the size of the tea leaves designated on a package of tea?

7. How did lemon juice affect the tea at room temperature? At refrigerator temperature?

Questions for Post-Lab Writing

8. Why does tea become less cloudy when lemon juice is added?

9. Why does tea change color when lemon juice is added?

10. What is the difference between "chocolate" and "cocoa"?

11. Which chocolate beverage looked the richest or darkest in color?

12. Which chocolate beverage tasted the richest?

13. Why did the commercial cocoa beverage have a reddish tint?

14. What caused the differences in mouthfeel between the chocolate and cocoa beverages?

15. Why is chocolate normally melted in a double boiler?

Appendix A – Objective Tests and Templates

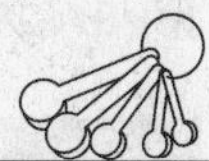

Objective Testing

Objective testing includes physical and chemical evaluations performed by laboratory instruments. Objective tests are designed to mimic sensory evaluations and provide quantitative data.

Linespread Test

The linespread test is a simple, reliable, objective evaluation that assesses the viscosity of a product based on its spreadability on a flat surface. It can be used to provide a comparison of the viscosity of similar products or to assess the influence of a particular ingredient in an identified food. The linespread test is standardized through time and temperature controls. Equipment required for the linespread test includes a diagram of concentric circles, a 2”-diameter hollow cylinder, and a spatula. The diagram of concentric circles can be copied on a transparency, and the transparency can be used in actual testing or a clear overlay or glass plate may be used over the template. Directions for the linespread test and a diagram of the concentric circles are included for testing.

Linespread Test
1. Position a glass or plastic plate over a diagram of concentric circles so that the center of the circles is directly under the center of the plate.
2. Place a hollow cylinder 2 inches in diameter with a volume of ¼ c. or 76 mL over the smallest circle and fill with the food to be tested (see Figure A-1).
3. Level the food in the cylinder with a spatula. Be sure there are no air pockets in the cylinder.
4. Lift the cylinder and allow the food to spread for exactly 30 seconds (see Figure A-2).
5. Quickly take four readings at the four numbered points on the circumference of the circles. Record the readings on the designated table.
6. Average the four readings. Record the average reading for each substance tested on the designated table.
7. Wash the glass plate and the hollow cylinder. Repeat the linespread test for additional substances to be tested, checking to be sure each test substance is at the specified temperature. Make sure the hollow cylinder and glass plate are room temperature.
8. A flow period of 2 minutes is often used rather than 30 seconds. The appropriate flow time will depend on the viscosity of specific food product. The flow time must be consistent for all samples.
9. A thinner-viscosity liquid will have a higher mean score.

Figure A-1: Place a 2" cylinder (biscuit cutter) on the linespread diagram and fill with substance.

Figure A-2: Lift off cylinder and allow substance to spread for the designated flow period before taking the linespread readings using the numbered circles.

Percent Sag Test

The percent sag test is another objective test that is used to measure gel strength. A skewer is used to measure the product height of a molded and an unmolded sample. The height difference is used to calculate the percent sag. A tender gel is expected to have a larger percent sag value, while a stronger gel has a smaller percent sag value. Specific directions to perform the sag test are provided.

Percent Sag Test
1. While the gel to be tested is still in the baking cup, insert a metal skewer into the center of the gel (see Figure A-3).
2. Remove the skewer and measure the depth of penetration to nearest millimeter (see Figure A-4).
3. Loosen the gel around the edges of the container with a metal spatula and carefully turn the gel onto a flat plate.
4. Immediately insert the skewer into the center of the gel (see Figure A-5). Remove the skewer and measure the depth of penetration to the nearest millimeter.
5. Calculate the percent sag for the custard. $$\% \text{ Sag} = \frac{\text{Height in container} - \text{Height out of container}}{\text{Height in container}} \times 100$$
Note: A larger percent sag value indicates a more tender gel or custard.

Figure A-3: Insert a metal skewer or glass rod into the gel within its container.

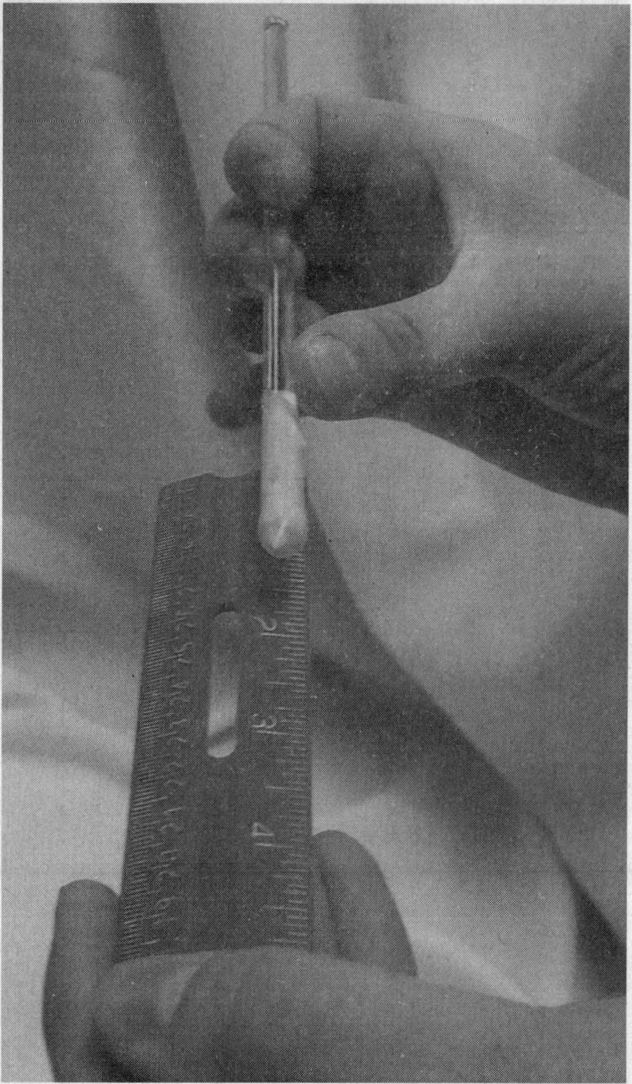

Figure A-4: Measure the depth of penetration into the gel using a ruler.

Figure A-5: Invert the gel onto a flat surface and insert a skewer or rod to measure depth of penetration (either immediately or after allowing it to sag for a designated period).

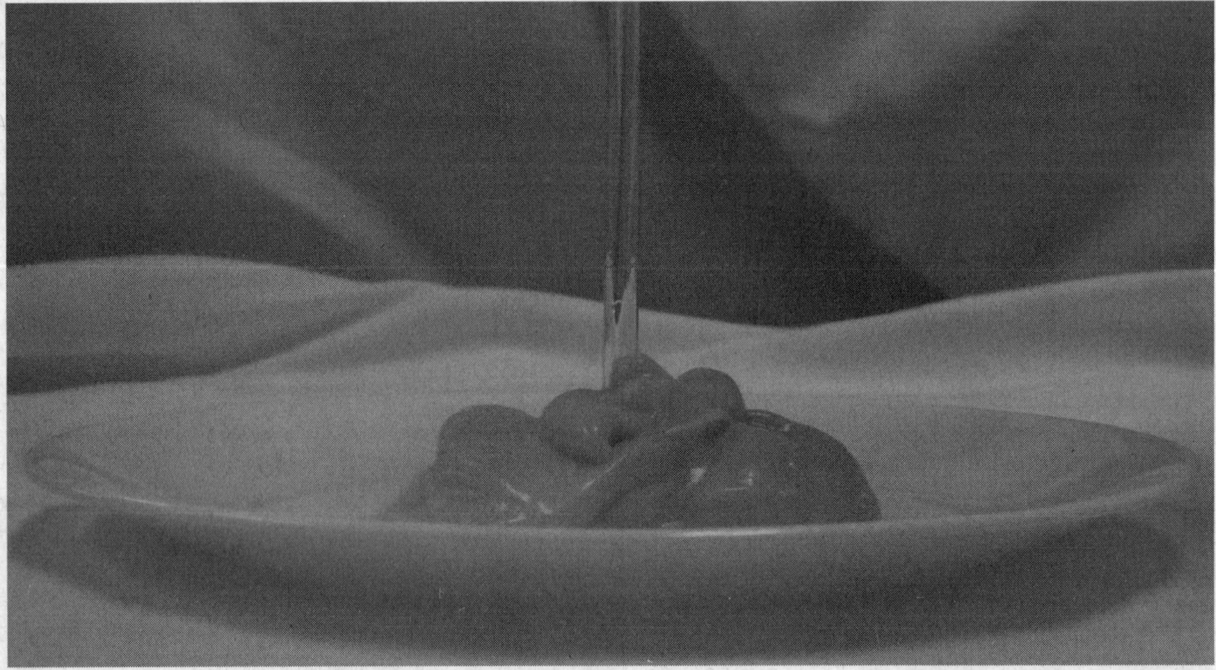

Templates

The linespread test template and "Cooking Record" form are included on the following pages for duplication and use during lab exercises.

The cooking record is a tool used to record product data in the meat lab. Percent yield is calculated as follows:

$$\% \text{ Yield} = \frac{\text{Cooked weight}}{\text{Original weight}} \times 100$$

Linespread Diagram

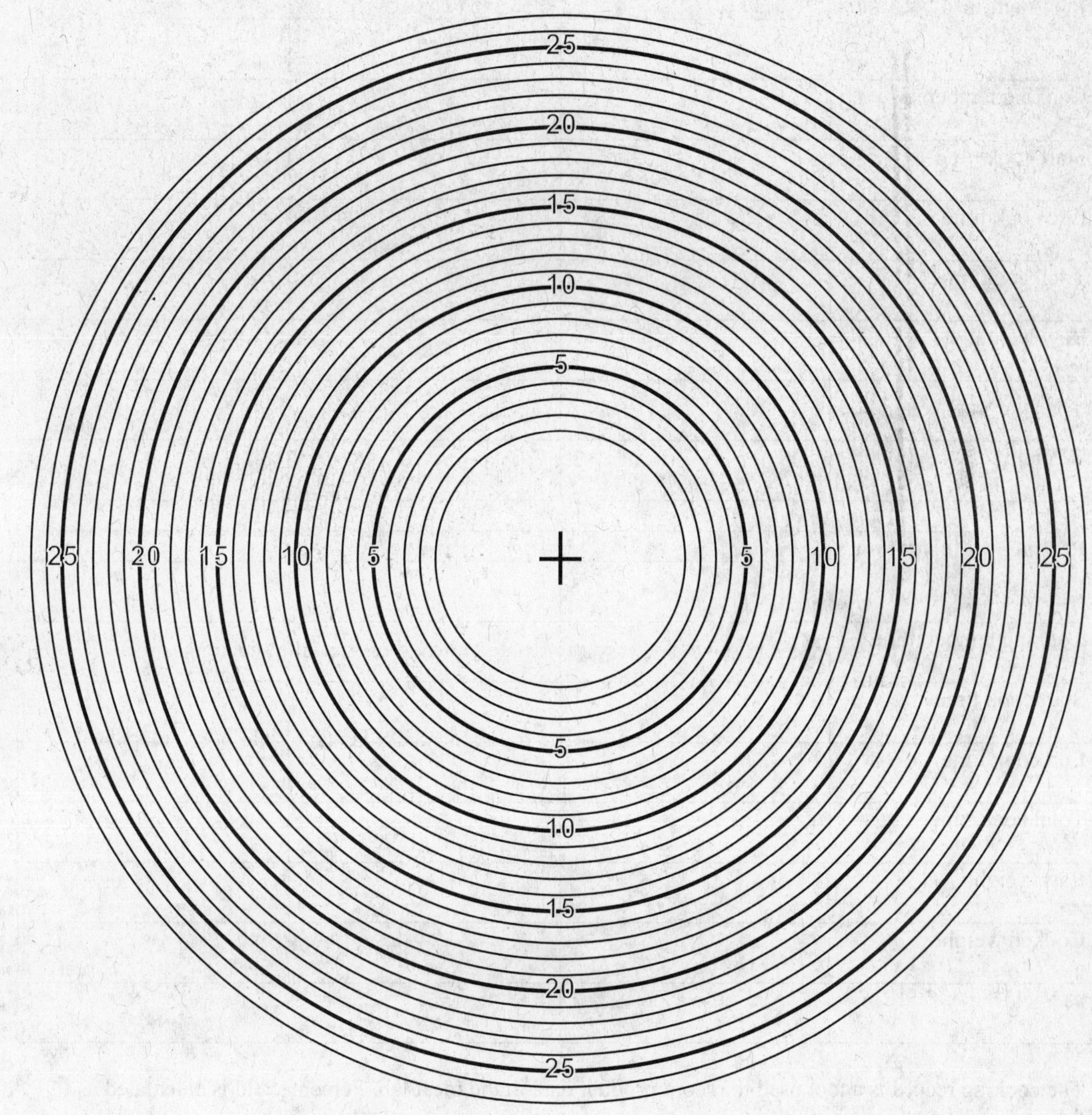

Cooking Record

Name/group ______________________________ Date ______________________

Pertinent information	Product		
Cooking temperature			
Start cook time			
End cook time			
Total cook time			
Raw weight			
Cooked weight			
% Yield*			

Pertinent information	Product		
Cooking temperature			
Start cook time			
End cook time			
Total cook time			
Raw weight			
Cooked weight			
% Yield*			

The cooking record is a tool used to record product data in the meat lab. Percent yield is calculated as follows:

$$\% \text{ Yield} = \frac{\text{Cooked weight}}{\text{Original weight}} \times 100$$

Appendix B – Instructor Guide: Suggestions, Materials Needed, and Estimated Completion Times

Food Lab Safety and Orientation

Here are some suggestions for activities to orient students to the lab environment and enhance safety awareness.

A. Scavenger Hunt

- Purpose is to familiarize students with laboratory and equipment used within the laboratory environment. Equipment required: see suggested list below.
- Scavenger hunt can be performed on an individual or group basis.
- Provide each group with a list of items that are located in the lab and have them either identify the location of these items (equipment) or gather the items (mixing utensils, graduated liquid measuring cups, etc.) in the lab.
- The first individual or first group to locate or find all of the items on their list wins the scavenger hunt.
- Suggested items included on the scavenger hunt list: hand-washing sink, 3-compartment sink, conventional oven, range, dishwasher, convection oven, scales, refrigerator, freezer, hand sanitizer, cleaning list, whisk, pastry cutter, graduated liquid measuring cup, candy thermometer, stem thermometer, digital thermometer, chlorine test strips, sanitizing water, clean dish towels, mixing bowls, sifter, apple corer, paring knife, cutting board, pH papers, egg separator, colander, etc.

B. Glow Potion Hand Washing Exercise

- Purpose is to promote good hand washing techniques. Equipment required: GlitterBug Potion[1], UV light.
- Dispense GlitterBug Potion on hands of all students in lab. Instruct students to cover hands with potion and rub in thoroughly.
- Instruct students to wash their hands. After students wash their hands, examine them with UV lamps. Bright spots indicate areas not thoroughly washed; these areas usually include hard-to-clean cracks and crevices between fingers, in nail beds, around rings, etc. After exposing hands to UV lights and identifying bright spots, send students to rewash hands until all bright spots are gone.

---

[1] Note: GlitterBug Potion is manufactured by the Brevis Corporation – www.brevis.com.

Unit 1 – Sensory Evaluation

A. Evaluation of Food Products Using Descriptive Terms

Ingredients Assortment of food products (chosen by instructor*)

*Suggestions: baked muffin; raw apple; strongly brewed black tea; lemon wedge; chocolate pudding

Equipment 2-oz. sample cups (number based on participants)
Serving utensil
Knife
Cutting board
Tea brewer or saucepan and range (if tea is used)

Estimated Prep Time: Depends on number of samples prepared; approximately 10 minutes for evaluation process.

B. Paired Comparison Test

Ingredients Two similar food or beverage products with assigned sample codes (chosen by instructor*)

*Suggestions: Use 2 sugar solutions with varying sweetness levels and colors:

- 13% sucrose solution (¼ c. sugar/ 2 c. water) – red food coloring
- 25% sucrose solution (½ c. sugar/ 2 c. water) – green food coloring

Equipment 2-oz. sample cups with assigned sample codes (number based on participants, 2 sample cups per participant)
Permanent marker (to mark sample cups)
Flask to prepare solutions
Stirring utensil
Measuring utensils
Graduated liquid measuring cup
Food coloring

Estimated Prep Time: Depends on number of samples prepared; approximately 5 minutes for evaluation process.

C. Triangle Test

Ingredients Two identical food or beverage products with assigned sample codes (chosen by instructor*)

One similar food or beverage product with an assigned sample code (chosen by instructor*)

*Suggestions: Use 2 red sports drinks such as Gatorade®, POWERade® or the store brand label; use Sprite® and 7UP®; use 2 cola beverages

Equipment 2-oz. sample cups with assigned sample codes (number based on participants, 3 sample cups per participant)
Permanent marker (to mark sample cups)

Estimated Prep Time: Depends on number of samples prepared; approximately 5 minutes for evaluation process.

D. Ranking Test

Ingredients 3-5 similar food or beverage products with varying intensity of specific characteristics and an assigned sample code (chosen by instructor*)

*Suggestions: mild cheddar cheese; medium-sharp cheddar cheese; and sharp cheddar cheese

Equipment 2-oz. sample cups with assigned sample codes (number based on participants, 3-5 sample cups per participant)
Permanent marker (to mark sample cups)
Knife (if cheese is used)
Cutting board (if cheese is used)

Estimated Prep Time: Depends on number of samples prepared; approximately 5 minutes for evaluation process.

E. Identification of Primary Tastes

Ingredients 5 coded samples that include each of the following:

- Tonic water
- 13% sucrose solution (¼ c. sugar/ 2 c. water)
- 6% vinegar solution (2 T. vinegar/ 2 c. water – 4% acetic acid)
- 4% NaCl solution (1.5 T. salt/ 2 c. water)
- Duplicate (1 of the previous 4 with a different sample code)

Unsalted crackers
Water

Equipment 2-oz. sample cups with assigned sample codes (number based on participants, 5 sample cups per participant)
Permanent marker (to mark sample cups)
4 flasks to mix solutions
Stirring utensils
Measuring utensils

Estimated Prep Time: Depends on number of samples prepared; approximately 5 minutes for evaluation process.

F-a. Spiced Rice

Ingredients 1 c. rice
1 t. margarine
Assortment of spices (chosen by instructor*)

*Suggestions: paprika; chili powder; curry powder; dry mustard; turmeric

Equipment Saucepan
Stirring utensil
Custard cups (number needed based on number of spices used)
Wax paper (or custard cups to display spices)
Range
Measuring utensils
Display labels
2-oz. sample cups (number based on participants)
Plastic spoons (number based on participants)

Estimated Prep Time: Depends on number of samples prepared; approximately 6-8 minutes for evaluation process.

F-b. Tomato Bouillon

Ingredients 1 c. tomato juice
2 c. beef bouillon (use cubes, liquid concentrate, or granules)
Assortment of herbs (chosen by instructor*)

*Suggestions: basil; cayenne; ground cumin; marjoram; oregano; parsley; poultry seasoning; rosemary; sage; tarragon; thyme

Equipment Saucepan
Range
Stirring utensil
Small teapots or measuring cups (number needed based on number of herbs used)
Measuring spoons
Graduated liquid measuring cup
Waxed paper
Display labels
2-oz. sample cups (number based on participants)

Estimated Prep Time: Depends on number of samples prepared; approximately 6-8 minutes for evaluation process.

F-c. Spiced Applesauce

Ingredients 1 Large jar applesauce
Assortment of spices (chosen by instructor*)

*Suggestions: ground coriander; allspice; ground ginger; mace; cardamom; nutmeg; cinnamon; pumpkin pie spice; ground cloves

Equipment Small bowls or custard cups (number needed based on number of herbs used)
Serving utensil
Stirring utensil
Measuring spoons
Waxed paper
Display labels
2-oz. sample cups (number based on participants)
Plastic spoons (number based on participants)

Estimated Prep Time: Depends on number of samples prepared; approximately 6-8 minutes for evaluation process.

F-d. Seasoned Cream Cheese

Ingredients 8 oz. cream cheese
1 ½ T. water
Assortment of herbs and seeds (chosen by instructor*)
Unsalted crackers

*Suggestions: dill weed; fennel seed; caraway seed; poppy seed; celery seed; savory; chervil; sesame seed

Equipment Mixing bowl
Electric mixer
Measuring spoons
Small bowls or custard cups (number needed based on number of herbs and seeds used)
Knife or spreader
Waxed paper
Display labels

Estimated Prep Time: Depends on number of samples prepared; approximately 6-8 minutes for evaluation process.

Unit 2 – Food Preparation Basics

B. Measuring Techniques for Flour

Ingredients ~4 c. flour

Equipment 1 c. measuring utensil
Sifter
Spoon
Scale
Spatula (straight edge)

C. Measuring Techniques for Sugar

Ingredients 1 c. granulated sugar
2 c. confectioners' sugar
2 c. brown sugar

Equipment 1 c. measuring utensil
Sifter
Spoon
Scale
Spatula (straight edge)

D. Measuring Techniques for Fats

Ingredients 2 c. canned hydrogenated shortening
1 c. stick hydrogenated shortening
1 c. stick margarine
1 c. whipped margarine
1 c. cold water

Equipment 1 c. metal measuring utensil
Knife
Cutting board
Spoon
Scale
3 plates for weighing product
Spatula (straight edge)
2 c. graduated liquid measuring cup

E. Measuring Techniques for Liquids

Ingredients Water

Equipment 1 c. graduated liquid glass measuring cup
Graduated cylinder

F. Using a Thermometer

Ingredients Water
Crushed ice

Equipment Stem thermometer that can be calibrated
Beaker
Pliers

G. Effect of Pan Surface Characteristics on Energy Transfer

Ingredients 1 ½ c. sugar
½ c. shortening
½ c. margarine
2 eggs
2 ¾ c. flour
2 t. cream of tartar
1 t. baking soda
¼ t. salt

Equipment Conventional oven
Convection oven
Mixing bowl
Electric mixer
Spoon
3 cookie sheets: very dark, light, air bake
Glass pan
Silicone pan
Cooling racks
Display plates
Display labels

H. Effect of Container Material on Energy Transfer

Ingredients Water

Equipment Saucepan
Stopwatch
Range
1 c. metal measuring cup
1 c. graduated Pyrex liquid measuring cup
Styrofoam cup
Thermometer with ring stand

I. Effect of Shape Container on Energy Transfer

Ingredients 4 c. water

Equipment Graduated liquid measuring cup
Small aluminum loaf pan
2 thermometers
Aluminum cake pan (8" × 8")
Oven with timer

J-a Microwave Chicken Preparation

Ingredients 2 t. margarine
2 chicken filets (4 oz. each)
¼ t. paprika
¼ t. salt
¼ t. pepper

Equipment Microwave-safe glass bowl
Microwave oven
Thermometer
Waxed paper
Display labels

J-b Microwave Biscuit Preparation

Ingredients 2 refrigerated canned biscuits

Equipment Microwave-safe plate
Microwave oven
Waxed paper
Knife
Cutting board
Display labels

K. Reheating Using Microwave Energy Transfer

Ingredients 8 baked dinner rolls

Equipment Microwave-safe plate
Microwave oven
Knife
Cutting board
Display labels

L. Defrosting Using Microwave Energy Transfer

Ingredients 2 frozen ground beef patties (¼ lb. each)

Equipment Microwave
2 microwave-safe plates
Display labels

Unit 3 – Meat

A. Comparison of Beef, Veal, Pork, and Lamb

Ingredients Analogous cuts (e.g., rib steak or chop) of beef, pork, veal, and lamb (chosen by instructor)

Equipment Oven with broiler and pans for broiling meat **or**
Electric grill
Tongs
Thermometer
Display plates (number based on samples)
Display labels (number based on samples)

Estimated Prep Time: Depends on number of samples prepared.

B. Comparison of Lean and Choice Beef

Ingredients 4 oz. analogous cuts of choice beef and lean/select beef

Equipment Oven with broiler and pans for broiling meat or
Electric grill
Tongs
Thermometer
Scale
2 display plates
Foil
Price information
Cooking Record (Appendix A)
Display labels
Holding oven

Estimated Prep Time: 8-10 minutes per variation

C. Comparison of Connective Tissue in Meat and Cooking Methods

Ingredients 4 oz. analogous cuts of brisket strip and loin strip

Equipment Oven with broiler and pans for broiling meat or
Electric grill
Tongs
Thermometer
Scale
2 display plates
Foil
Cooking Record (Appendix A)
Display labels
Holding oven

Estimated Prep Time: 13-15 minutes per variation

D. Effect of Tenderizers on Meat

Ingredients 5 pieces of top round steak (4 oz. each)
½ t. commercial meat tenderizer
½ c. vinegar
¼ t. salt
1 fresh pineapple

Equipment Meat hammer or jacquard
Measuring spoons
Food processor
Knife
Cutting board
Oven with broiler and pans for broiling meat **or**
Electric grill
Tongs
Thermometer
Scale
5 treatment plates
5 display plates
Display labels
Foil
Cooking Record (Appendix A)
Holding oven

Estimated Prep Time: 8-10 minutes per variation

E. Comparison of Beef Patty Products

Ingredients ¼ lb. 73% ground beef patty
¼ lb. ground turkey patty
¼ lb. 90% lean ground sirloin patty
Boca® burger patty
¼ lb. beef patty containing soy protein
¼ lb. fully cooked hamburger patty

Equipment Electric skillet
Thermometer
Cooking Record (Appendix A)
6 plates
6 display plates
Foil
Price information on patties
Display labels
Scale
Spatula

Estimated Prep Time: 14-16 minutes per variation

F. Comparison of Protein Products Used in Stir Fry

Ingredients 3 T. oil, divided
½ lb. beef for stir-fry
½ lb. pork
½ lb. firm tofu
3 c. frozen stir-fry vegetables, divided
6 T. Worcestershire sauce, divided
½ c. soy sauce, divided

Equipment Wok or electric skillet
Knife
Cutting board
Wooden stirring utensils
Measuring utensils

Price information for beef, tofu, and lamb
3 display plates
Display labels

Estimated Prep Time: 20 minutes per variation

G-a. Swiss Steak

Ingredients
1 lb. round steak
2 T. flour
2 T. butter or margarine
1 clove garlic
½ t. salt
¼ t. dry mustard
Pepper to taste
½ t. Worcestershire sauce
½ c. tomato juice
Chopped vegetables (e.g., onion, celery, carrot, green pepper, mushrooms) to taste
2 T. flour (optional, for gravy)

Equipment
Knife
Cutting board
Meat hammer or jacquard
Plastic or paper bag
Measuring utensils
Electric skillet with lid **or** oven with timer, oven proof casserole dish, and foil
Tongs
Whisk

Estimated Prep Time: 2.5 hours

G-b. Breaded Veal Cutlets

Ingredients
1 slice veal cutlet
1 egg
¼ c. + 1 T. Water
½ c. bread crumbs
½ t. salt
Pepper to taste
2 T. fat

Equipment
Knife
Cutting board
2 mixing bowls
Measuring utensils
Electric skillet with lid
Spatula
Whisk

Estimated Prep Time: 1 hour

G-c. Oven-Braised Pork Chops

Ingredients
½ lb. pork chops
2 T. flour
¼ t. salt
$^{1}/_{8}$ t. pepper
1 egg, beaten
1 T. milk
$^{1}/_{3}$ c. bread crumbs
2 T. oil

Equipment
Knife
Cutting board
Plastic/paper bag
Mixing bowl
Waxed paper
Skillet
Range
Oven-proof casserole dish
Oven with timer
Tongs

Estimated Prep Time: 45- 50 minutes

G-d. Glazed Pork Tenderloin

Ingredients
2-3 lb. pork tenderloin
½ t. salt
$^{1}/_{8}$ t. pepper
4 T. honey
2 T. brown sugar
2 T. cider vinegar
1 t. spicy brown mustard

Equipment
Roasting pan with rack
Foil
Oven
Bowl
Measuring spoons
Large spoon
Meat thermometer

Estimated Prep Time: 2.25 hours

G-e. Pot Roast

Ingredients
2-3 lb. beef pot roast
1 T. vegetable oil
1 envelope dried onion soup mix
2 cans (10.5 oz) cream of mushroom soup
Water
Aerosol vegetable oil cooking spray

Equipment
Skillet
Range
Can opener
Large meat fork
9” × 13” roasting pan
Foil
Bowl
Spoon
Oven
Meat thermometer

Estimated Prep Time: 4.25 hours

Unit 4 – Poultry

A. Comparison of Generic and National Brand Chicken

Ingredients
2 store brand boneless chicken breast halves
2 name brand boneless chicken breast halves*
2 eggs, beaten, divided
1 c. dry bread crumbs, divided
4 T. vegetable oil, divided
*Suggestions – Tyson®, Pilgrim's®, Foster Farms®

Equipment
1 bowl
1 plate or waxed paper
Electric skillet or range and skillet
Whisk
Tongs
Thermometer
Alternative: broiler pan and oven
Display labels

Estimated Prep Time: 20-25 minutes per variation

B. Turkey Deli Meat Comparison

Ingredients
Least expensive deli turkey with an assigned sample code*
Mid-priced deli turkey with an assigned sample code*
Premium priced deli turkey with an assigned sample code*
*Quantity needed based on number of participants

Equipment
2-oz. sample cups with assigned sample codes (number based on participants, 3 sample cups per participant)
Permanent marker (to mark sample cups)
Knife
Cutting board
Cost information

Estimated Prep Time: Depends on number of samples prepared; approximately 10 minutes for evaluation process.

C-a. Chinese Almond Chicken

Ingredients
½ c. slivered almonds
2 uncooked chicken breast halves, skinless and boneless
1 T. oil
1 c. sliced onion
1 ½ c. diagonally-cut celery pieces
1 c. chicken broth
1 T. cornstarch
1 t. granulated sugar
¼ c. soy sauce
¾ c. chicken broth
5 oz. canned bamboo shoots, drained
5 oz. canned water chestnuts, drained and sliced
½ lb. fresh broccoli, cut into florets
¼ lb. fresh or frozen pea pods

Equipment
Knife
Cutting board
Skillet
Stirring utensil
Range
Measuring utensils
2 bowls

Estimated Prep Time: 35-40 minutes

C-b. Chicken Divine

Ingredients
*20 oz. frozen broccoli spears
1 lb. cooked chicken breast, deboned, bite size pieces
2 10.5-oz. cans cream of mushroom soup
1 c. mayonnaise
1 t. lemon juice
½ t. curry powder
2 T. cooking sherry
1 c. medium cheddar cheese, shredded
30 round butter flavored crackers, crushed
Aerosol vegetable oil cooking spray

*May reduce fat content by substituting equivalent amounts of low-fat cream of mushroom soup, low-fat mayonnaise, and saltine crackers for cream of mushroom soup, mayonnaise, and butter crackers.

Equipment
Saucepan with lid (for broccoli preparation)
Saucepan with lid or baking pan (for chicken preparation)
Range
Mixing bowl
Mixing utensil
Measuring utensils
Knife
Cutting board
Can opener
Colander
Cheese grater
Plastic bag
11 ½" × 7 ½" × 2" baking pan.
Oven with timer
Thermometer

Estimated Prep Time: 45-50 minutes

C-c. Glazed Cornish Hen

Ingredients
1 Cornish hen (1-1 ½ pounds)
1 T. butter
2 T. apricot jam or orange marmalade
Small amount of orange or cranberry juice
Dash dry mustard
Salt
Pepper

Equipment
Paper towels
Roasting pan with rack
Foil
Oven
Bowl
Measuring utensils
Stirring utensil
Basting utensil
Meat thermometer

Estimated Prep Time: 1 hour, 33 minutes

C-d. Turkey and Broccoli Burgers

Ingredients
10 oz. package frozen chopped broccoli
1 ½ lbs. lean ground turkey
1 c. extra sharp cheddar cheese, shredded
1 small onion, chopped
1 egg, beaten
¼ c. dried bread crumbs
1 T. Worcestershire sauce
1 t. salt
Aerosol vegetable oil cooking spray

Equipment
Saucepan with lid
Range
Knife
Cutting board
Colander
Large mixing bowl
Mixing utensil
Measuring utensils
Whisk
Cheese grater
Baking pan
Oven
Spatula
Thermometer

Estimated Prep Time: 45 minutes

C-e. Chicken Spaghetti

Ingredients
Whole chicken
1 T. salt
Water
2 celery stalks, diced
1 medium onion, diced
1 green bell pepper, diced
Garlic clove, minced
1 T. chili powder
½ t. oregano
10.5 oz cream of mushroom soup
2 cans tomato sauce (8 oz. each)
1 c. medium cheddar cheese, grated
13.25 oz. spaghetti
Aerosol vegetable oil cooking spray

Equipment
2 large pots with lid
Thermometer
Pan (for cooling)
Can opener
Measuring utensils
Graduated liquid measuring cup
Large spoon
Knife
Cutting board
Cheese grater
9” × 13” baking dish
Oven

Estimated Prep Time: 1.5 hours

C-f. Chicken Salad

Ingredients
4 chicken breasts **or** 1 whole chicken
Water
¼ c. celery, chopped
1 c. seedless red grapes, halved
1 can mandarin oranges
1 small can crushed pineapple
1 red apple cut up
½ c. slivered almonds
½ c. mayonnaise
¼ c. poppy seed dressing

Equipment
Large pot with lid
Range
Thermometer
Pan for cooling
Mixing bowl
Measuring utensils
Graduated liquid measuring cup
Knife
Cutting board
Bowl
Stirring utensil
Refrigerator

Estimated Prep Time: 1.5 hours

Unit 5 – Fish and Shellfish

A. Coagulation of Fish Protein by Heat—Comparison of Fresh and Frozen Fish

Ingredients
- 12 oz. fresh flounder (or other white fish) fillets, divided
- 12 oz. thawed, frozen flounder (or other white fish) fillets, divided
- 2 T. melted butter, divided
- Aerosol vegetable oil cooking spray
- 1 T. + 1 t. salt, divided
- Water

Equipment
- Broiler with rack
- Foil
- Brush (for butter)
- Measuring utensils
- Pot
- Range
- Cheesecloth and string
- Microwave-safe baking dish/ casserole with lid
- Microwave oven
- 6 plates for display
- Display labels

Estimated Prep Time: 10 minutes per poached variation; 20 minutes per grilled variation

B-a. Shrimp Bourgeoise

Ingredients
- 1 lb. fresh or frozen shrimp
- 2 T. butter
- ½ clove garlic, minced
- 1 T. chopped fresh parsley
- $^1/_8$ t. salt
- Dash of cayenne pepper and black pepper
- 1 T. dry white wine
- ½ c. canned stewed tomatoes

Equipment
- Saucepan
- Range
- Measuring utensils
- Stirring utensil
- Knife
- Cutting board

Estimated Prep Time: 20 minutes

B-b. Surimi Creole

Ingredients
- 1 T. minced onion
- 1 T. minced green pepper
- 8-10 mushrooms, sliced
- 2 T. fat or oil
- 2 c. canned tomatoes, chopped
- ½ t. chili powder or pepper
- ½ t. salt
- ¾ lb. surimi, cut into bite-size pieces
- ½ c. uncooked rice
- Water

Equipment
- 1 saucepan with lid
- Saucepan
- Stirring utensil
- Range
- Measuring utensils
- Knife
- Cutting board
- Graduated liquid measuring cup

Estimated Prep Time: 30 minutes

B-c. Surimi Cocktail

Ingredients
- 1 lb. surimi, cut into bite-size pieces
- 2 c. seafood cocktail sauce

Equipment
- Plate
- Toothpicks

Estimated Prep Time: 5 minute

B-d Crayfish Cornbread

Ingredients
- 2 eggs
- 1 t. salt
- 1 medium onion, chopped
- ½ c. vegetable oil
- 1 c. yellow corn meal
- 1 c. medium cheddar cheese, shredded
- ½ fresh jalapeno, chopped
- 16 oz. cream corn
- 1 lb. crayfish tails
- Aerosol vegetable oil cooking spray

Equipment
- Knife
- Cutting board
- Cheese shredder
- Large mixing bowl
- Measuring utensils
- Stirring utensil
- Can opener
- 2 qt. rectangular baking pan
- Oven with timer
- Thermometer

Estimated Prep Time: 1 hour

B-e. Oven-Roasted Salmon with Caper Sauce

Ingredients
- 1½ lb. salmon fillet
- Salt and freshly ground black pepper
- Extra virgin olive oil
- ¼ c. Hellmann's® real mayonnaise **or** sour cream
- ¼ c. nonfat plain yogurt
- 1 T. freshly squeezed lemon juice
- ¼ t. Tabasco sauce or to taste
- ¼ - ½ t. Worcestershire sauce
- 3 T. brine-cured capers, rinsed and chopped
- 1 t. Dijon mustard

Equipment Mixing bowl
Whisk
Baking pan
Oven with timer
Measuring utensils
Knife
Cutting board
Foil

Estimated Prep Time: 45 minutes

Unit 6 – Milk

A. Sampling of Milk Products
Ingredients Assortment of milk products (chosen by instructor*)
*Suggestions: whole milk; lactose free milk; 2% milk; evaporated milk; skim milk: soy milk; reconstituted nonfat dry milk; organic milk; buttermilk
Equipment 2-oz. sampling cups (number needed based on participants and samples)
Package label
Price information
Display labels

Estimated Prep Time: Depends on number of samples prepared; approximately 10 minutes for evaluation process.

B. Sampling of Yogurt Products
Ingredients Assortment of yogurt products (chosen by instructor*)
*Suggestions: yogurt with various fat contents; yogurt with probiotics: live bacteria, no-bacteria gels
Equipment 2-oz. sampling cups (number needed based on participants and samples)
Package label
Price information
Display labels

Estimated Prep Time: Depends on number of samples prepared; approximately 10 minutes for evaluation process.

C. Coagulation of Fresh Whole Milk
Ingredients 1 ½ c. whole milk, divided
2 T. vinegar, divided
Equipment Saucepan
8 strips pH paper
Range
Measuring spoons
Graduated liquid measuring cup

Estimated Prep Time: 15 minutes

D. Preparation of Scalloped Potatoes – Effect of Tannins on Milk Proteins
Ingredients 4 c. Raw, peeled, thinly sliced potatoes, divided
2 c. whole milk, divided
2 T. flour
2 T. butter
½ t. salt
Aerosol vegetable oil cooking spray
Equipment 2 1-qt casserole dishes
Range
Saucepan
Stirring utensil
Measuring utensils
Graduated liquid measuring cup
Oven with timer
Display labels

Estimated Prep Time: 45 minutes

E. Vanilla Pudding
Ingredients 1 c. granulated sugar, divided
9 T. cornstarch, divided
$^3/_8$ t. salt, divided
3 T. butter, divided
1 T. vanilla extract, divided
2 c. whole milk, divided
2 c. soy milk
1 c. 2% fat milk
1 8-oz. cup low-fat vanilla or fruited yogurt
Equipment Saucepan per variation (3 variations)
Stirring utensils
Measuring utensils
Range
3 serving dishes
Refrigerator
Display labels

Estimated Prep Time: 45 minutes per variation

F. Milk Foams
Ingredients 1 c. heavy whipping cream, chilled, divided
½ c. heavy whipping cream, room temperature
1 c. evaporated milk, frozen with ice crystals, divided
2 t. lemon juice
½ c. reconstituted non-fat milk solids
Equipment Thermometer
Refrigerator/freezer
6 mixing bowls (5 chilled, 1 room temperature)
6 electric mixers (5 sets of beaters chilled, 1 room temperature)

Glass graduated measuring cup
6 funnel liners
6 funnels
6 graduated cylinders
Freezer tray
Scale
Display labels

Estimated Prep Time: 10-15 minutes most variations; 30 minutes for butter

Unit 7 – Cheese

A. Cheese Display

Ingredients Assortment of cheeses* (chosen by instructor)
Unsalted crackers*
Apples*

*Quantity based on number of samples and participants

Equipment Knife
Cutting board
Apple corer
Tray for apples
Plate for unsalted crackers
Refrigerator
Display plates for cheese
Display labels for each cheese variety

B. Effect of Heat on Cheeses

Ingredients 14 saltine crackers
1 slice ($^1/_8$" × 1" × 3") of each of the following:

- mild cheddar cheese
- medium cheddar cheese
- sharp cheddar cheese
- extra sharp cheddar cheese
- process cheese
- process cheese food
- process cheese spread

Equipment Baking sheet
Aluminum foil
Oven
Cooling rack
Spoon or fork
Labels for each cheese variety
Display plates

C. Preparation of Cottage Cheese

Ingredients 4 c. skim milk, divided
1 rennet tablet
Water
Commercial dry curd cottage cheese (16 oz. or smaller tub)
2 T. vinegar + extra if needed

Equipment 2 small saucepans
Measuring utensils
Utensil for slicing curd
3 strips pH paper
2 skillets
White paper
Range
Small bowl
Thermometer
Scale
2 cheesecloth bags
Graduated cylinder
3 display containers
Display labels

D. Cheese Sauces

Ingredients 3 c. milk, divided
4 T. butter, divided
4 T. flour, divided
1 t. salt, divided
3 oz. natural medium sharp Cheddar cheese, shredded
3 oz. process cheese, shredded
Water
2 t. baking soda, divided
2 t. cream of tartar, divided

Equipment 3 small saucepans per variation (2 variations)
Range
Stirring utensils
Double boiler
6 strips pH paper
Measuring utensils
Graduated liquid measuring cup
6 display containers
Display labels

E-a. Chicken Cheese Soup

Ingredients Whole chicken
Water
¼ c. margarine
½ c. onion, diced
½ c. carrots, grated
½ c. celery, diced
¼ c. flour
1 ½ T. cornstarch
4 c. chicken broth (from boiled chicken)
4 c. milk
$^1/_8$ t. baking soda
1 lb. Ole English cheese, cubed
1 t. salt
¼ t. pepper
1 T. dried parsley

Equipment Large pot with lid
Thermometer
Pan for cooling

Heavy saucepan
Large spoon
Range
Measuring utensils
Graduated liquid measuring cup
Knife
Cutting board

E-b. Cheese Ball

Ingredients 16 oz. cream cheese
2.5 oz. Buddig beef, chopped
½ bunch green onions, chopped
¼ c. chopped pecans

Equipment Electric mixer
Spatula
Mixing bowl
Measuring utensils
Knife
Cutting board
Waxed paper
Saran® wrap
Refrigerator

Unit 8 – Eggs

A. Quality of Raw Eggs

Ingredients 1 fresh grade A egg
1 deteriorated grade A egg (egg which has been held at room temperature for at least one week)

Equipment 2 flat plates
Display labels
Price per dozen

Estimated Prep Time: 5 minutes

B. Hard-Cooked Fresh and Deteriorated Eggs

Ingredients 4 c. water
2 fresh eggs
2 deteriorated eggs (eggs which have been held at room temperature for at least one week)

Equipment 2 small saucepans
Range
Thermometer
Timer
Knife
4 display plates
Display labels

Estimated Prep Time: Up to 1 hour

C. Scrambled Eggs

Ingredients 2 fresh eggs, divided
3 T. milk, divided
$^{3}/_{16}$ t. salt, divided
1 T. butter, divided
Egg substitute (amount equivalent to 1 egg)

Equipment Fork
Mixing bowl per variation (2 variations)
Measuring utensils
Small frying pan
Range
Spatula
Microwave
Small microwave-safe dish
3 display plates
Display labels

Estimated Prep Time: 10 minutes

D. Comparison of Baked and Stirred Custards

Ingredients 2 c. milk, divided
2 eggs, divided
4 T. granulated sugar, divided
Water

Equipment Electric mixer
Oven
Double boiler
3 custard cups
Range
Square metal pan
Thermometer
Metal skewer
Stirring utensil
Linespread diagram
Stopwatch
Spatula
Transparent plate
Ruler
2” diameter hollow cylinder
Flat plate
Display labels
Sink

Estimated Prep Time: 30 minutes per variation

E. Factors Influencing Egg White Foams

Ingredients 6 egg whites, separated (1 for each of 6 egg foam variations)
2 T. granulated sugar
¼ t. cream of tartar
2 T. Splenda
½ t. egg yolk

Equipment Egg separator
Electric mixer
Mixing bowl per variation (6 variations)
Stopwatch
Glass graduated measuring cup
6 funnel liners

6 glass funnels
6 100-mL graduated cylinders
Display labels

Estimated Prep Time: 10-15 minutes per variation

F. Preparation of Angelfood Cake Using Egg Foams

Ingredients 2 c. cake flour, sifted before measuring
3 c. granulated sugar
1 t. salt
3 t. cream of tartar
3 t. vanilla
1 ½ c. room temperature egg whites (8-10 large eggs)
1 ½ c. egg substitute

Equipment Sifter
2 mixing bowls
Measuring utensils
Egg separator
Electric mixer
Rubber spatula
2 tube pans
Sheet pan
Oven with timer
Display labels

Estimated Prep Time: 1.25 hours

G. Preparation of Brownie Products Using Egg Substitutes

Ingredients 5 packages brownie mix
1 or more eggs, as required for 1 package mix
1 egg white substitute per egg required for 1 package mix
¼ c. soft tofu per egg required for 1 package mix
1 T. milled flax seed per egg required for 1 package mix
Water
1 ½ T. oil per egg required for 1 package mix
1 t. baking powder per egg required for 1 package mix
Additional (non-egg) ingredients required to prepare 5 packages mix
Aerosol vegetable oil cooking spray

Equipment Mixing bowl per variation (5 variations)
Measuring utensils
Spatula
5 baking pans (8” × 8”)
Oven with timer

Estimated Prep Time: 30 minutes per variation

H. Meringues

Ingredients 2 egg whites, separated (1 for each of 2 meringue variations)
6 T. sugar, divided

Equipment Electric mixer
Measuring spoons
Egg separator
Spatula
Mixing bowl per variation (2 variations)
2 baking sheets
2 ovens with timers

Estimated Prep Time: Soft meringue – 20 minutes; hard meringue – 1.25 hours

I-a. Cheese Soufflé

Ingredients ¼ c. butter
¼ t. salt
1 ½ c. milk
¼ c. flour
Dash of black pepper
8 oz. shredded sharp cheddar cheese
2 T chopped, fresh parsley
6 eggs, separated
½ t. cream of tartar

Equipment Oven with timer
Medium saucepan
Measuring utensils
Range
Stirring utensil
Small bowl
Large bowl
Electric mixer
Egg separator
2-quart casserole or soufflé dish

Estimated Prep Time: 1.75 hours

I-b. Ham Quiche

Ingredients 1 pastry for 9” pie
1 ½ c. (6 oz.) shredded Swiss Cheese
1 T. flour
½ c. diced cooked ham
4 eggs, beaten
2 c. half and half
½ t. dry mustard
¼ t. salt

Equipment 9” pie pan
Knife
Fork
Oven with timer
2 bowls
Measuring utensils
Electric mixer

Estimated Prep Time: 1.5 hours

I-c. Deviled Eggs

Ingredients ½ dozen eggs
Water
¼ c. salad dressing
1 t. prepared mustard
½ t. salt
Dash pepper
½ T. dill pickle relish
Paprika

Equipment Saucepan
Thermometer
2 spoons
Mixing bowl
Electric mixer
Measuring utensils
Graduated liquid measuring cup
Timer
Knife
Cutting board
Serving platter
Refrigerator

Estimated Prep Time: 45 minutes

Unit 9 – Vegetables and Fruits

A. Enzymatic Oxidative Browning

Ingredients 1 tart apple (e.g., Granny Smith)
1 sweet apple (e.g., Red Delicious)
1 potato
Iceberg lettuce leaves
Water
Pineapple juice
Lemon juice
Acid solution (dissolve ¼ t. ascorbic acid in ½ c. water)
Dry ascorbic acid
Cream of tarter solution (dissolve ¼ t. cream of tarter in ¼ c. water)
Sucrose solution (dissolve 2 T. sugar in ¼ c. water)

Equipment Knife
Cutting board
Measuring utensils
40 custard cups
Saucepan
Range
4 strips pH paper
Plastic wrap
Refrigerator
Display labels

Estimated Prep Time: 30 minutes set-up; 15 minutes for evaluation

B. Effect of Cooking Medium on Apples

Ingredients 8 apples of the same variety, divided
8 apples of a second variety, divided
10 c. water, divided
3 c. sugar, divided
2 ½ c. sugar substitute (Splenda®)

Equipment Knife or apple corer
Cutting board
Saucepan per variation (8 variations)
Range
Measuring utensils
Graduated liquid measuring cup
Slotted spoon
8 display plates
Display labels

Estimated Prep Time: 35 minutes

C. Effects of Cooking Treatments on Vegetable Pigments

Ingredients 1 ½ lb. broccoli, divided
1 ½ lb. carrots, divided
1 ½ lb. red cabbage, divided
1 ½ lb. cauliflower, divided
Water
8 t. cream of tartar, divided
2 t. baking soda, divided

Equipment Knife
Cutting board
Measuring utensils
Slotted spoons
5 stainless steel saucepans with lids per variation (4 variations)
Steamer basket per variation (4 variations)
Range
Small microwave safe dish with lid per variation (4 variations)
Microwave
24 custard cups
24 strips pH paper
White plates
Display labels

Estimated Prep Time: 45 minutes per variation

D. Factors That Affect Chlorophyll

Ingredients ½ lb. fresh green beans
Water
16 oz. canned green beans
10 oz. frozen green beans

Equipment Skillet with lid
2 saucepans with lids
Can opener
Range
Slotted spoon

3 white plates
2 custard cups
Display labels

Estimated Prep Time: 15 minutes

E. Preparation of Applesauce

Ingredients 3 delicious apples
3 Jonathan or Winesap apples
3 apples of either variety
2 ¼ c. water, divided
¾ c. sugar, divided
3 sprinkles cinnamon, divided

Equipment Knife
Cutting board
3 1-qt. saucepans with lids
Range
Dry measuring cup
Liquid measuring cup
3 forks

Estimated Prep Time: 15 minutes

F-a. Fruit Crumble

Ingredients 2 c. fresh fruit (blueberries, blackberries, raspberries, apples, peaches, pears, or a mixture)
¼ c. sugar (or less, depending on the sweetness of the fruit)
1 T. flour
Pinch of salt
½ t. grated lemon rind
1¼ c. vanilla wafer crumbs
¼ c. 2 T. flour
¼ t. cinnamon
¼ c. melted butter

Equipment 2 bowls
Measuring utensils
Stirring utensils
Grater
Casserole dish
Oven with timer

Estimated Prep Time: 50 minutes

F-b. Guacamole

Ingredients 2 ripe avocados, mashed
1 ripe tomato, peeled and finely chopped
1 T. fresh lime or lemon juice
½ t. salt
Tabasco® sauce to taste
2 T. chopped onion
4 T. canned green chilies, finely chopped
Tortilla chips

Equipment Knife
Cutting board
Can opener
Measuring utensils
Mashing utensil
Stirring utensil
Mixing bowl
Blender (optional)

Estimated Prep Time: 25 minutes

G-a. Artichoke with Lemon-Butter Sauce

Ingredients 1 artichoke
½ lemon, sliced thinly
¼ c. butter
2 t. lemon juice
½ t. salt
Water for cooking artichoke

Equipment Scissors
Knife
Cutting board
Measuring utensils
Slotted spoon
Saucepan with lid
Range
Mixing bowl
Mixing utensil

Estimated Prep Time: 1 hour

G-b. Braised Carrots and Leeks with Yogurt-Dill Sauce

Ingredients 8 medium carrots, peeled and cut into ¼-inch slices
1 bunch leeks
2 T. butter
Salt and pepper to taste
1 c. plain yogurt
2 T. chopped fresh dill (or dried dill to taste)
¼ c. chicken bouillon

Equipment Knife
Cutting board
Vegetable peeler
Measuring utensils
Graduated liquid measuring cup
Frying pan
Range
Bowl
2 stirring utensils

Estimated Prep Time: 35-40 minutes

G-c. Spaghetti Squash with Tomato Sauce

Ingredients 1 spaghetti squash
1 onion, chopped

1 clove garlic, crushed
15 oz. canned stewed tomatoes
12 oz. tomato sauce
1 t. basil
1 t. oregano
2 T. olive oil
1 t. salt
1 ½ t. sugar
Pepper to taste

Equipment Knife
Cutting board
Can opener
Measuring utensils
Large, deep saucepan with lid
Skillet
Stirring utensil
Range
Spoon
Serving dish

Estimated Prep Time: 30 minutes

G-d. Steamed Parsnips

Ingredients 6 fresh parsnips, peeled and thinly sliced
3 T. butter
Salt and pepper

Equipment Knife
Cutting board
Slotted spoon
Measuring utensils
Saucepan
Range

Estimated Prep Time: 30 minutes

G-e. Squash Casserole

Ingredients 5 fresh yellow squash, sliced
1 c. water
1 medium onion, diced
½ t. salt
Dash pepper
Envelope cornbread mix (ingredients for cornbread)
1 t. sage
10.5 oz. cream of chicken soup
1 c. Velveeta® cheese, shredded
Aerosol vegetable oil cooking spray

Equipment Knife
Cutting board
Graduated liquid measuring cup
Measuring utensils
Saucepan with lid
Range
Large mixing bowl
Mixing utensil
Oven with timer
Can opener
Cheese grater
9" × 13" baking pan

Estimated Prep Time: 45-50 minutes

G-f. Glazed Carrots

Ingredients 16 oz peeled baby carrots
⅓ c. apricot jam
¼ c. brown sugar
½ t. ground cinnamon
1 T. butter

Equipment Oven with timer
Foil
Baking sheet
Measuring utensils
Mixing bowl
Stirring utensil

Estimated Prep Time: 30 minutes

G-g. Broccoli Casserole

Ingredients 1 26 oz. frozen broccoli florets
Water
1 c. mayonnaise
1 c. Colby Jack cheese, grated
1 can (10.5 oz) Cream of chicken soup
4 oz. cream cheese
2 eggs, lightly beaten
1 sleeve butter crackers, crushed
2 T. butter, melted
Aerosol vegetable oil cooking spray

Equipment Oven with timer
9 × 13" baking dish
Saucepan with a lid
Stirring utensil
Can opener
Measuring utensils
Whisk
2 mixing bowls
Colander
Plastic bag to crush crackers

Estimated Prep Time: 50 minutes

Unit 10 – Legumes

A. Preparing Dried Legumes

Ingredients 2 c. dried legumes (chosen by instructor)
8 c. water
2 t. cream of tarter
½ t. baking soda
Additional water for cooking

Equipment 3 bowls
Colander
Measuring utensils
Stirring utensil
4 saucepans with lids
Range
pH papers
Display containers
Display labels

Estimated Prep Time: 10 hours for soaking beans – 9 hours soaking and 1 hour cooking

B-a. Lentil Salad

Ingredients 1 c. dried lentils
1 onion stuck with 2 cloves
1 bay leaf
3 c. water
1 t. salt
2 ½ T. oil
1 ½ T. wine vinegar
Finely chopped scallions, to taste
2 T. fresh chopped parsley
Ground black pepper, to taste

Equipment Saucepan
Knife
Cutting board
Measuring utensils
Graduated liquid measuring cup
Range
Colander
Stirring utensil
Mixing bowl

Estimated Prep Time: 1 hour

B-b. White Beans with Tomatoes and Garlic

Ingredients 3 c. canned cannelloni or other white beans
¼ c. olive oil
1 t. garlic, minced
½ t. dried sweet basil
2 large ripe tomatoes, peeled and coarsely chopped
½ t. salt
Dash ground black pepper
1 T. wine vinegar

Equipment Measuring utensils
Graduated liquid measuring cup
Knife
Cutting board
Large colander
Heavy 10-inch skillet with lid
Stirring utensil
Range

Estimated Prep Time: 20 minutes

B-c. Herbed Soy Bean Patties

Ingredients 16 oz. canned soybeans, drained
$^1/_8$ t. garlic powder
½ t. dried thyme
3 T. wheat germ
½ t. salt
Fresh whole-wheat bread crumbs
½ medium onion, chopped finely
¼ - ½ t. rosemary, to taste
½ c. fresh parsley, chopped
1 large egg, beaten
$^1/_8$ t. pepper

Equipment Measuring utensils
Mashing utensil
Can opener
Stirring utensil
Knife
Cutting board
Mixing bowl
Small mixing bowl
Whisk
Shallow pan for bread crumb breading
Shallow baking dish
Oven

Estimated Prep Time: 45 minutes

B-d. Chili with Corn

Ingredients 3 T. oil
1 onion, finely chopped
1 clove garlic, cut in half
1 green pepper, chopped
2 c. vegetable stock or water
1 c. coarsely chopped, peeled tomato
2 T. tomato paste
1 c. cooked corn
4 c. cooked kidney or pinto beans
½ t. chili powder
¼ t. cumin powder
1 ½ t. salt
1 t. oregano
Ground pepper, to taste

Equipment Knife
Cutting board
Measuring utensils
Graduated liquid measuring cup
Skillet with lid
Range
Bowl
Mashing utensil
Stirring utensil
Can opener

Estimated Prep Time: 45-50 minutes

B-e. Falafel in Pita Bread

Ingredients
16 oz. canned garbanzo beans (chickpeas), drained
1 slice whole wheat bread
2 T. chopped fresh parsley
1 T. tahini (sesame seed paste)
½ t. turmeric
2 eggs, lightly beaten
½ t. salt
⅛ t. cayenne pepper
Whole-wheat pita bread
Lettuce, tomato, tahini or yogurt as garnish

Equipment
Blender
Can opener
Knife
Cutting board
Measuring utensils
Frying pan
Small mixing bowl
Whisk
Spatula
Range
Paper towels

Estimated Prep Time: 45 minutes

B-f. Chickpea Appetizer (Hummus)

Ingredients
10 oz. (1 ¼ c.) canned chickpeas
1 ½ T. sesame tahini mixed with 1 ½ T. cold water
2 T. lemon juice
½ clove garlic, minced
½ t. salt
¼ t. black pepper

Equipment
Colander
Blender
Can opener
Knife
Cutting board
Measuring utensils
Stirring utensil
Serving bowl

Estimated Prep Time: 20 minutes

B-g. Tofu Fruit Smoothie

Ingredients
1 carton soft tofu
1 c. frozen raspberries, blueberries, or strawberries (or a combination)
2 T. honey (adjust to taste)
½ c. fruit juice

Equipment
Blender
Measuring utensils
Graduated liquid measuring cup
Stirring utensil
Serving utensil
Refrigerator

Estimated Prep Time: 10 minutes

B-h. Soy Milk Fruit Smoothie

Ingredients
1 c. soy milk (plain or vanilla)
½ medium banana (may be frozen if desired)
2 T. frozen fruit juice concentrate, undiluted
¼ c. frozen fruit such as peaches or berries

Equipment
Blender
Graduated liquid measuring cup
Measuring utensils
Stirring utensil
Serving utensil
Refrigerator

Estimated Prep Time: 10 minutes

B-i. Tofu Broccoli Quiche

Ingredients
1 pre-made pie crust, pre-cooked for 12 minutes or until lightly brown
1 lb. broccoli, trimmed and cut into small pieces (or substitute chopped spinach)
1 medium onion, finely chopped
2 cloves garlic, minced
2 T. oil
1 lb. firm tofu, drained
½ c. milk or soy milk
¼ t. Dijon mustard
¾ t. salt
¼ t. ground nutmeg
½ t. ground red pepper
2 T. Parmesan cheese or soy Parmesan cheese

Equipment
Saucepan with lid
Range
Skillet
Stirring utensil
Measuring utensils
Graduated liquid measuring cup
Knife
Cutting board
Blender
Large bowl
Oven with timer

Estimated Prep Time: 1.25 minutes

B-j. Mocha Mousse

Ingredients
2 ½ c. mashed silken firm tofu
½ c. sugar

$^{1}/_{3}$ c. unsweetened cocoa powder
1 t. orange zest
2 T. freeze-dried instant coffee granules
½ t. ground cinnamon

Equipment Food processor or blender
Refrigerator
Measuring utensils
Spoon
Serving dishes

Estimated Prep Time: 1.25 hours

Unit 11 – Cereals and Flours

A. Market Forms of Oatmeal Cereals

Ingredients 1 serving regular rolled oats
2 servings quick-cooking oats
1 serving instant oatmeal
Water
Salt

Equipment 2 saucepans with lids
1 saucepan
Bowl
Measuring utensils
Graduated liquid measuring cup
250-mL graduated cylinder
Stirring utensils
Range
4 display containers
Display labels

Estimated Prep Time: 1-20 minutes depending on product

B. Comparison of Types of Rice

Ingredients 1 c. uncooked brown rice
1 c. uncooked long-grain white rice
1 c. uncooked short-grain white rice
1 c. uncooked converted rice
1 c. uncooked instant rice
Water, divided
2 t. salt, divided
4 T. oil, divided

Equipment 4 1-qt. saucepans
Graduated liquid measuring cup
250-mL graduated cylinder
Range
4 wooden spoons
Measuring utensils
4 colanders
4 display bowls
Display labels

Estimated Prep Time: 5-50 minutes depending on product

C. Comparison of Types of Pasta

Ingredients ½ c. uncooked elbow macaroni
½ c. uncooked egg noodles
½ c. uncooked spaghetti
½ c. uncooked whole-wheat pasta
8 c. water, divided

Equipment 1 Saucepan with lid per variation (5 variations)
Graduated liquid measuring cup
Range
Stirring utensil
Measuring utensils
5 display containers
Display labels

Estimated Prep Time: 15 minutes per variation

D. Preparation of Gluten Ball

Ingredients 1 c. bread flour
1 c. all-purpose flour
1 c. cake flour
1 c. whole wheat flour
1 c. rye flour
1 c. cornmeal flour
1 ½ c. water
Fresh water for rinsing starch

Equipment 1 mixing bowl per variation (6 variations)
Sink
Mixing utensil
Graduated liquid measuring cup
Scale
Waxed paper
Cookie sheet
Oven with timer
6 display plates
Display labels

Estimated Prep Time: 1.25-1.5 hours depending on product

E-a. Hominy Grits

Ingredients Package of hominy grits
Water

Equipment Saucepan with lid
Range
Stirring utensil
Graduated liquid measuring cup
Measuring utensils

Estimated Prep Time: 20 minutes

E-a. Tortillas with Chili Con Queso

Ingredients **Tortillas:**
2 $^{1}/_{3}$ c. instant masa harina (corn flour)
1 t. salt
1 - 1½ c. cold water

Chili Con Queso Sauce:
3 T. butter
3 T. flour
8-oz. can tomatoes with Jalapeno peppers
4-oz. can green chilies, drained and chopped
6 oz. Monterey Jack cheese, shredded
¼ - ½ c. milk

Equipment Deep bowl
Graduated liquid measuring cup
Measuring utensils
Range
2 stirring utensils
Towel
Skillet
Oven
Tongs
Saucepan
Can opener
Double boiler

Estimated Prep Time: 45 minutes

E-b. Couscous

Ingredients 1 package couscous
Water
1 T. melted butter

Equipment Saucepan with lid
Graduated liquid measuring cup
Stirring utensil
Measuring utensils
Range
Fork

Estimated Prep Time: 20 minutes

E-b. Farina

Ingredients 1 package farina
Water

Equipment Saucepan with lid
Range
Stirring utensil
Measuring utensils
Graduated liquid measuring cup

Estimated Prep Time: 20 minutes

E-b. Almond Poppy Seed Noodles

Ingredients 6 oz. spinach noodles
¼ c. sliced almonds
1½ T. poppy seeds
2-3 T. butter

Equipment Saucepan
Measuring utensils
Stirring utensil
Pot
Colander
Utensil for tossing
Range
Serving dish

Estimated Prep Time: 30 minutes

E-b. Tabouleh

Ingredients 2 tomatoes, chopped
1 cucumber, peeled and chopped
½ c. finely chopped fresh parsley
½ c. chopped scallions
2 c. bulgur (cracked wheat)
2 T. olive oil
2 T. lemon juice
1 t. dried mint leaves (optional)
Salt and pepper to taste
Pita bread or romaine lettuce leaves

Equipment Knife
Cutting board
Measuring utensils
Bowl
Colander
Stirring utensil

Estimated Prep Time: 45 minutes

E-b. Orzo Salad With Tomatoes and Herbs

Ingredients 8 oz. orzo pasta (about 1¼ c.)
Water
3 T. sherry wine vinegar
1 T. fresh lemon juice
½ C. extra-virgin olive oil
1 ¼ lb. cherry tomatoes
¾ c. green onions or chives, chopped
½ c. pitted oil-cured olives or pitted Kalamata olives, sliced
¼ c. fresh basil, thinly sliced
¼ c. fresh mint, chopped
¼ c. fresh Italian parsley, chopped
Salt and pepper to taste

Equipment Pot
Range
Sink
Colander
Measuring utensils
Graduated liquid measuring cup
Knife
Cutting board
2 mixing bowls (one medium, one small)
Whisk
Stirring utensil

Estimated Prep Time: 45 minutes

E-c. Buckwheat Pancakes

Ingredients
1 c. buckwheat flour
1 c. all-purpose flour
½ t. salt
2 T. brown sugar
4 t. baking powder
3 T. oil
2 eggs, beaten
1 ¼ - 1 ½ c. milk

Equipment
Sifter
2 bowls
Measuring utensils
Stirring utensil
Graduated liquid measuring cup
Griddle or frying pan
Spatula
Range
Pan
Oven

Estimated Prep Time: 15 minutes

E-c. Kasha

Ingredients
1 c. kasha (buckwheat groats)
1 egg, beaten
1 t. salt
4 T. butter
½ c. chopped onions
¼ lb. chopped fresh mushrooms
2-2 ½ c. boiling water

Equipment
Knife
Cutting board
Mixing bowl
Whisk
Measuring utensils
Graduated liquid measuring cup
Stirring utensils
Skillet with lid
Saucepan
Spatula
Range
Sauté pan

Estimated Prep Time: 45 minutes

E-d. Brown Rice with Almonds, Onions and Cheese

Ingredients
1 c. brown rice
2 c. boiling water
½ t. salt
½ onion, chopped
3 T. sliced almonds
2 T. butter
$^{1}/_{3}$ c. shredded cheddar cheese

Equipment
Pot with lid
Measuring utensils
Knife
Cutting board
Range
Stirring utensils
Sauté pan

Estimated Prep Time: 60 minutes

E-e. Oatmeal Muffins

Ingredients
1 ¼ c. quick cooking rolled oats
1 ¼ c. milk
1 ¼ c. sifted all purpose flour
4 t. baking powder
½ t. salt
1 egg, slightly beaten
½ c. oil
$^{1}/_{3}$ c. firmly packed brown sugar
½ c. raisins (optional)

Equipment
2 small bowls
Graduated liquid measuring cup
Measuring utensils
1 large mixing bowl
Sifter
Stirring utensils
6 cup muffin pan
Muffin liners
Oven with timer

Estimated Prep Time: 1 hour

E-e Cowboy Cookies

Ingredients
1 c. sugar
1 c. brown sugar
1 c. shortening
2 c. flour
1 t. baking soda
½ t. salt
½ t. baking powder
2 eggs
2 c. oatmeal
1 t. vanilla extract
1 ½ c. chopped pecans
9 oz. chocolate chips
1 ½ c. coconut

Equipment
Large mixing bowl
Small mixing bowl
Electric mixer
Spatula (with straight edge)
Measuring utensils
Sifter
Cookie Sheet
Oven with timer

Estimated Prep Time: 45 minutes

E-e. Peanut Butter Chocolate Granola Bar

Ingredients
½ c. brown sugar
1/3 c. butter
1/3 c. crunchy peanut butter
1/3 c. honey
1 egg
½ t. vanilla extract
1 c. chocolate chips
3 ½ c. quick oatmeal
¼ c. pecans (optional)
Aerosol vegetable oil cooking spray

Equipment
Large mixing bowl
Electric mixer
Measuring utensils
Graduated liquid measuring cup
Spatula (with straight edge)
8" × 8" baking pan
Oven with timer

Estimated Prep Time: 45 minutes

E-f. Vegetable Barley Soup

Ingredients
¼ onion, chopped
2 carrots, diced
1 stalk celery, diced
1 turnip, diced
½ c. green beans, sliced
½ c. whole barley, uncooked
2 T. oil
1 qt. hot water
1 t. salt
Pepper to taste
¼ t. marjoram*
¼ t. thyme*
1 T. chopped fresh parsley

*Substitute any herbs of your choice.

Equipment
Knife
Cutting board
Measuring utensils
Sauté pan
Graduated liquid measuring cup
Pot with lid
Range
Stirring utensil

Estimated Prep Time: 60 minutes

E-f. Barley Pilaf

Ingredients
¾ c. barley, uncooked
2½ c. beef broth
1/3 c. chopped onion
¼ c. chopped green pepper
¼ c. sliced celery
1 T. butter

Equipment
Pot with lid
Measuring utensils
Graduated liquid measuring cup
Range
Stirring utensil
Sauté pan

Estimated Prep Time: 50 minutes

E-g. Gluten-Free Banana Muffins

Ingredients
1 c. rice flour
1 c. sweet sorghum flour
1 t. baking soda
½ t. baking powder
¼ t. nutmeg
¾ T. cinnamon
1 egg
1 ½ c. apple juice
2 ripe bananas (mashed)
¼ c. honey
1/3 c. brown sugar
½ c. raisins

Equipment
Oven with timer
Measuring utensils
Stirring utensils
Mixing bowl
Muffin pans
Cupcake liners
Graduated liquid measuring cup
Toothpick

Estimated Prep Time: 30 minutes

E-g. Gluten-Free Apple Muffins

Ingredients
1 c. rice flour
1 c. sweet sorghum flour
1 t. baking soda
½ t. baking powder
¼ t. nutmeg
¾ T. cinnamon
1 egg
1 ½ c. apple juice
1 medium apple, peeled, cored and diced
¼ c. honey
1/3 c. brown sugar
½ c raisins

Equipment
Oven with timer
Measuring utensils
Stirring utensils
Mixing bowl
Muffin pans
Cupcake liners
Graduated liquid measuring cup
Toothpick
Paring knife
Cutting board

Estimated Prep Time: 30 minutes

E-g. Gluten-Free Maple Cookies

Ingredients
1 c. vegetable shortening
½ c. pure maple syrup
½ c. brown sugar
2 eggs
3 c. sweet sorghum flour
½ t. baking soda
1 t. salt
1 ½ t. gluten-free (and corn syrup free) vanilla
½ t. cinnamon

Equipment
Oven with timer
Measuring utensils
Spatula
Mixing bowl
Electric mixer
Sifter
Cookie sheet
Waxed paper
Refrigerator
Graduated liquid measuring cup
Knife
Cutting board

Estimated Prep Time: 45 minutes

Unit 12 – Starches and Sauces

A. Vanilla Cornstarch Puddings

Ingredients
3 T. cornstarch
¼ c. + 2 T. granulated sugar
2 c. milk + milk required for preparing cooked and instant pudding mixes, divided
$^{1}/_{8}$ t. salt
1 t. vanilla extract
1 package cooked vanilla pudding mix
1 box instant vanilla pudding mix
1 can high-amylopectin vanilla pudding
6 T. flour

Equipment
Measuring utensils
Graduated liquid measuring cup
2 Saucepans
Mixing bowl
Stirring utensils
Timer
Range
~ 12 custard cups with 4 lids or plastic wrap
Refrigerator
Display labels

Estimated Prep Time: 20 minutes

B. Effect of Starch Variety on Lemon Pie Filling

Ingredients
2 T. cornstarch
4 T. all-purpose flour
4 T. quick-cooking tapioca
2 T. potato starch
3 T. arrowroot starch
2 ½ c. granulated sugar, divided
1 ¼ c. cold water, divided
3 ¾ c. boiling water, divided
5 egg yolk, beaten, divided
¾ c.+ ½ T. lemon juice, divided
$^{5}/_{16}$ t. salt, divided
2 ½ t. grated lemon rind, divided
5 T. butter, divided

Equipment
Small bowl per variation (5 variations)
Whisk
Grater
Graduated liquid measuring cup
Measuring utensils
2-qt. saucepan per variation (5 variations)
Stirring utensil
Range
Bowl
5 custard cups
Display labels

Estimated Prep Time: 25-35 minutes per variation

C. Effect of Type of Starch on Viscosity of Starch Pastes

Ingredients
2 T. waxy corn starch
2 T. cornstarch
2 T. rice starch
2 T. quick tapioca
2 T. potato starch
2 T. arrowroot starch
4 T. all-purpose flour, divided
2 T. whole-wheat flour
6 ¾ c. water, divided
Ice water (for cooling)

Equipment
1–pt. heavy sauce pan per variation (9 variations)
Measuring utensils
Graduated liquid measuring cup
Double boiler
Stirring utensil
Skillet (for dextrinized flour)
Range
Thermometer
Linespread test materials per variation – linespread diagram, clear plate, 2" diameter hollow cylinder, spatula (9 variations)
18 custard cups with lids or plastic wrap
Pan

Freezer
Display labels

Estimated Prep Time: 20-60 minutes depending on product

D. Effect of Acid and Sugar on Viscosity of Starch Pastes

Ingredients 8 T. cornstarch, divided
¾ c. sugar, divided
8 T. lemon juice, divided
3 ½ c. water, divided
Ice water (for cooling)

Equipment 1-qt. heavy saucepan per variation (4 variations)
Measuring utensils
Graduated liquid measuring cup
Stirring utensils
Range
Thermometer
pH papers
Linespread test materials per variation – linespread diagram, clear plate, 2" diameter hollow cylinder, spatula (4 variations)
4 custard cups with lids or plastic wrap
Display labels
Pan

Estimated Prep Time: 25-35 minutes per variation

E. White Sauces

Ingredients 10 T. flour, divided
8 T. butter, divided
2 t. salt, divided
4 c. milk, divided

Equipment Saucepan per variation (4 variations)
Range
Stirring utensil
Thermometer
Measuring utensils
Graduated liquid measuring cup
Linespread test materials per variation – linespread diagram, clear plate, 2" diameter hollow cylinder, spatula (4 variations)
Display labels

Estimated Prep Time: 30 minutes per variation

F. Banana Pudding

Ingredients*Large (5.25 oz.) instant vanilla pudding
2 c. milk
½ t. vanilla extract
⅔ c. sugar
8 oz. sour cream
8 oz. Cool Whip®
Vanilla wafers
4-6 large bananas

*May reduce fat and caloric content by substituting equivalent amounts of sugar-free, fat-free instant vanilla pudding, skim milk, Splenda®, fat-free sour cream, low-fat Cool Whip® and reduced-fat vanilla wafers for regular vanilla pudding, whole milk, sugar, sour cream, Cool Whip® and vanilla wafers.

Equipment Electric mixer
Large mixing bowl
Spatula
Measuring utensils
Graduated liquid measuring cup
Knife
Cutting board
9" × 9" Pyrex dish

Estimated Prep Time: 25 minutes

Unit 13 – Quick Breads

A. Chemical Leaveners

Ingredients 6 c. flour, divided
1 ½ t. salt, divided
2 c. whole milk, divided
1 c. buttermilk, divided
9 large eggs, divided
12 T. oil, divided
6 T. sugar, divided
1 ½ t. soda, divided
1 T. double-acting baking powder, divided

Equipment 12 mixing bowls
12 wooden spoons
Measuring utensils
Graduated dry measuring cup
Graduated liquid measuring cup
Measuring utensils
10 muffin tins
Double oven
Cooling rack
Serrated knife
Cutting board
Ruler
10 display plates

Estimated Prep Time: 25 minutes per variation

B. Leavening by Air

Ingredients 1 c. cake flour
1 ½ c. sugar
1 ½ c. egg whites

¼ t. salt
1 ¼ t. cream of tartar
1 t. vanilla

Equipment Sifter
2 mixing bowls
Electric mixer
Measuring utensils
Rubber spatula
3 loaf pans (9¼" × 5¼" × 2¾")
Oven with timer
Cooling rack
3 display plates
Display labels

Estimated Prep Time: 1 hour

C. Effect of Egg Protein on Popovers

Ingredients 2 c. flour, divided
2 c. milk, divided
3 eggs
½ t. salt, divided
2 T. melted butter, divided
Aerosol vegetable oil cooking spray

Equipment Mixing bowls
Measuring utensils
Electric mixer
Muffin tins
Oven with timer
2 display plates
Display labels

Estimated Prep Time: 1 hour

D. Effect of Manipulation and Flour Type on Muffins

Ingredients 2 c. all-purpose flour
1 ½ t. salt, divided
7 ½ t. double-acting baking powder, divided
6 T. sugar, divided
6 T. oil, divided
3 eggs
3 c. milk, divided
2 c. cake flour
2 c. whole-wheat flour
Aerosol vegetable oil cooking spray

Equipment 2 mixing bowls per variation (3 variations)
Measuring utensils
Graduated liquid measuring cup
Sifter
Stirring utensils
36 muffin liners (optional)
3 muffin pans (12 tins each)
Oven with timer
36 display plates
Display labels

Estimated Prep Time: 30 minutes per variation

E. Effect of Manipulation on Biscuits

Ingredients 3 c. flour, divided
¾ t. salt, divided
4 ½ t. double-acting baking powder, divided
6 T. shortening, divided
15-21 T. milk, divided
1 package commercial biscuit mix
Ingredients needed to prepare mix (see package)

Equipment Sifter
Mixing bowl per variation (4 variations)
Measuring utensils
Pastry cutter
Stirring utensil
Fork
Kneading board
Biscuit cutter
Baking sheets
Oven with timer
4 display plates
Display labels

Estimated Prep Time: 30 minutes per variation

F. Effect of Sugar on Corn Bread

Ingredients 2 c. yellow or white corn meal, divided
2 c. flour, divided
¼ c. granulated sugar
8 t. double-acting baking powder, divided
1 t. salt, divided
2 c. milk, divided
2 eggs, divided
½ c. oil, divided
Aerosol vegetable oil cooking spray

Equipment Large mixing bowl per variation (2 variations)
Smaller mixing bowl per variation (2 variations)
Measuring utensils
Graduated liquid measuring cup
Stirring utensils
2 baking pans (8" × 8")
Oven with timer
2 display Plates
Display labels

Estimated Prep Time: 35 minutes per variation

G. Leavening by Steam: Cream Puff Shells

Ingredients 1 c. water
½ c. butter
¼ t. salt

1 c. flour
4 eggs
Aerosol vegetable oil cooking spray

Equipment Saucepan
Measuring utensils
Graduated liquid measuring cup
Range
Wooden spoon
Baking sheet
Oven with timer

Estimated Prep Time: 1.25 hours

H-a. Banana Bread

Ingredients 1 c. all-purpose flour
¾ c. whole-wheat flour
¼ c. wheat germ
¼ c. roasted sunflower seeds (optional)
1 T. double-acting baking powder
⅓ c. milk
1 c. mashed ripe bananas
2 eggs, slightly beaten
½ c. honey
4 T. oil
Aerosol vegetable oil cooking spray

Equipment 2 Large mixing bowls
Measuring utensils
Graduated liquid measuring cup
Stirring utensil
9" × 5" loaf pan
Oven with timer
Toothpick

Estimated Prep Time: 1.25 hours

H-b. Raisin-Bran Muffins

Ingredients 1 c. all-purpose flour
1 c. bran cereal
2 t. double-acting baking powder
1 egg, slightly beaten
2 T. oil
2 T. honey
¼ c. molasses
⅓ c. raisins
1 c. buttermilk
Aerosol vegetable oil cooking spray

Equipment 3 bowls
Measuring utensils
Sifter
Whisk
Stirring utensil
6 cup muffin tin
Oven with timer

Estimated Prep Time: 45 minutes

H-c. Cranberry Pumpkin Bread

Ingredients 2 eggs
¼ c. vegetable oil
1 ½ c. sugar
8 oz. canned pumpkin
1 ¾ c. + 2 T. flour
1 t. baking soda
½ t. salt
½ t. cinnamon
¼ t. ground nutmeg
⅛ t. ground cloves
⅛ t. ground ginger
1 c. whole fresh cranberries
½ c. chopped pecans (optional)
Aerosol vegetable oil cooking spray

Equipment Large mixing bowl.
Small mixing bowl
Electric mixer
Can opener
Spatula (with straight edge)
Measuring utensils
Graduated liquid measuring cup
9" × 5" loaf pan
Oven with timer
Toothpick
Cooling rack

Estimated Prep Time: 1.5 hours

H-d. Date Nut Bread

Ingredients 8 oz. dates
1 t. baking soda
1 c. boiling water
1 egg, beaten
1 c. sugar
2 c. + 2 T. flour, divided
1 t. baking powder
½ t. salt
1 t. vanilla extract
½ c. chopped pecans
Shortening

Equipment Large mixing bowl
Small mixing bowl
Saucepan
Range
Spatula (with straight edge)
Measuring utensils
Graduated liquid measuring cup
Sifter
Whisk (to beat egg)
Bowl (to beat egg)
9" × 5" loaf pan
Oven
Toothpick
Paper towel (to grease pan)

Estimated Prep Time: 1.5 hours

H-e. Pancakes

Ingredients
1 ¼ c. flour
½ t. salt
1 T. baking powder
1 T. sugar
1 egg
1 c. milk
2 T. oil

Equipment
Sifter
2 bowls
Measuring utensils
Stirring utensil
Graduated liquid measuring cup
Griddle or frying pan
Spatula
Range
Pan
Oven

Estimated Prep Time: 15 minutes

Unit 14 – Yeast Breads

A. Preparation of Yeast Breads by Various Methods

Ingredients
5 ½ t. salt, divided
7 envelopes active dry yeast, divided
2 c. + 2 T. water, divided
10 ¾ c. bread flour, approximately, divided
3 ¼ c. milk, divided
7 T. granulated sugar, divided
7 $^{1}/_{3}$ T. shortening, divided
Aerosol vegetable oil cooking spray
Ingredients required by the bread maker's manufacturer's recipe
1 T. melted butter

Equipment
1 saucepan per variation (7 variations)
1 small bowl per variation (7 variations)
1 large mixing bowl per variation (7 variations)
Thermometer
Range
Oven with timer
Graduated liquid measuring cup
Stirring utensils
Measuring utensils
Sifter
Waxed paper **or** 6 towels
Bread maker with manufacturer recipe
Muffin tin
6 loaf pans (9¼" × 5¼" × 2¾")
Kneading board
Proof box or warm area, 80 °F for rising
7 display plates
Display labels

Estimated Prep Time: 1.5-2 hours

B-a. Easy Yeast Rolls

Ingredients
3 ½ to 4 ½ c. flour, unsifted
3 T. sugar
1 t. salt
2 packages active dry yeast
1 c. milk
½ c. water
¼ c. butter or margarine
Aerosol vegetable oil cooking spray

Equipment
Mixing bowl
Towel or waxed paper
Saucepan
Thermometer
Baking pan
Range
Electric mixer
Spatula
Kneading board
Oven with timer

Estimated Prep Time: 1.5 hours

B-b. Whole-Wheat Bread

Ingredients
½ t. salt
1 envelope active dry yeast
2 T. water
$^{7}/_{8}$ c. all-purpose flour, approximately
$^{7}/_{8}$ c. whole-wheat flour, approximately
½ c. milk
2 t. granulated sugar
2 t. shortening
Aerosol vegetable oil cooking spray

Equipment
Mixing bowl
Loaf pan
Saucepan
Measuring utensils
Graduated liquid measuring cup
Proof box or warm area, 80 °F for rising
Oven with timer
Range
Stirring utensil
Kneading board or surface
Towel or waxed paper

Estimated Prep Time: 1.5-2 hours

B-c. Rye Bread

Ingredients
½ t. salt
1 envelope active dry yeast
2 T. water
$^{7}/_{8}$ c. all-purpose flour, approximately

$^{7}/_{8}$ c. rye flour, approximately
1-2 T. caraway seeds
½ c. milk
2 t. granulated sugar
2 t. shortening
Aerosol vegetable oil cooking spray

Equipment Mixing bowl
Loaf pan
Saucepan
Measuring utensils
Graduated liquid measuring cup
Proof box or warm area, 80 °F for rising
Oven with timer
Range
Stirring utensil
Kneading board or surface
Towel or wax paper

Estimated Prep Time: 1.5-2 hours

B-d. Cinnamon Rolls

Ingredients ½ t. salt
1 envelope active dry yeast
2 T. water
1 ¾ c. all-purpose flour, approximately
½ c. + 2 T. milk, divided
4 T. granulated sugar, divided
2 T. shortening
1 t. melted butter
1 t. cinnamon
2 T. nuts
2 T. raisins
1 c. confectioners' sugar, sifted
Aerosol vegetable oil cooking spray

Equipment Mixing bowl
Loaf pan
Saucepan
Measuring utensils
Graduated liquid measuring cup
Proof box or warm area, 80 °F for rising
Oven with timer
Range
Stirring utensil
Sifter
Kneading board or surface
Towel or waxed paper
Knife
Baking pan
Pastry brush
Small bowl
Whisk

Estimated Prep Time: 2 hours

B-e. Pita Bread

Ingredients ½ envelope active dry yeast
$^{2}/_{3}$ c. lukewarm water
½ T. honey
½ t. salt
2 c. all-purpose flour
1 T. oil

Equipment Mixing bowl
Measuring utensils
Graduated liquid measuring cup
Aluminum foil
Wooden spoon
Kneading surface
Rolling pin
Towel or waxed paper
Oven with timer
Baking sheet

Estimated Prep Time: 2.5 hours

Unit 15 – Fats and Oils

A. Comparison of Selected Commercial Fat Products Used as Spreads

Ingredients Assortment of spreads (as chosen by instructor*)
Unsalted crackers (quantity dependant on number of participants and samples)

*Suggestions: Shedd's Spread Country Crock® whipped margarine; Shedd's Spread Country Crock® stick margarine; Brummel & Brown® spread; Land O' Lakes whipped butter; Land O' Lakes stick butter; and Smart Balance® low-sodium spread.

Equipment Refrigerator
Butter knives (one per sample)
Display labels
Spread package label including ingredients and Nutritional analysis
Spread cost information

Estimated Prep Time: Depends on number of samples and participants; 10-15 minutes for evaluation

B. Effect of Coating on Fat Absorption

Ingredients 2 medium-sized raw yellow squash, divided
½ c. flour
½ t. salt
¼ t. pepper
1 egg
2 c. soybean oil

Equipment Knife
Cutting board
Measuring utensils

Graduated liquid measuring cup
2 bowls
Stirring utensil
Whisk
Thermometer
Deep-fat fryer with basket (may substitute electric skillet or skillet and range)
Paper towels
2 display plates
Display labels

Estimated Prep Time: 20 minutes per variation

C. Effect of Cooking Time on Fat Absorption

Ingredients 8 refrigerated canned biscuits, divided
4 c. soybean oil, divided

Equipment Knife
Cutting board
2 deep-fat fryers with baskets (may substitute electric skillets or skillets and range)
Paper towels
8 display plates
Display labels
Waxed paper or plates for weighing
Scale
Timer

Estimated Prep Time: 30-40 minutes

D. Stir-Fry and Oven Fry Methods

Ingredients 2 T. oil
4 medium zucchini, divided
1 small onion, chopped
¼ t. barlic powder, divided
½ t. salt, divided
1 large egg
½ c. grated parmesan cheese
½ c. fine breadcrumbs
Aerosol vegetable oil cooking spray

Equipment Vegetable brush
Sink
Knife
Cutting board
Skillet
Range
Stirring utensil
2 mixing bowls
Whisk
Baking pan
Spatula
Oven with timer
2 display plates
Display labels

Estimated Prep Time: 10-12 minutes per variation

E. Polyunsaturated versus Monounsaturated Oils: Vegetable Oil and Extra Virgin Olive Oil

Ingredients 1 lb. cucumbers, divided
1 lb. tomatoes, divided
1 red onion, cut in half and divided
1 t. salt, divided
2 pinches black pepper, divided
½ c. vegetable oil
¼ c. vinegar, divided
½ c. extra virgin olive oil

Equipment Knife
Cutting board
Large bowl per variation (2 variations)
Mixing bowl per variation (2 variations)
Stirring utensil
2 display plates
Display labels

Estimated Prep Time: 30 minutes

F. Fat and Fat Substitutes

Ingredients ⅓ c. melted margarine
⅓ c. applesauce
⅓ c. baby food prunes
¾ c. cocoa, divided
3 c. sugar, divided
6 eggs, divided
2 c. flour, divided
2 t. vanilla extract, divided
Aerosol vegetable oil cooking spray

Equipment Oven with timer
3 baking pans (8” × 8”)
Small saucepan
Range
Large mixing bowl per variation (3 variations)
Measuring utensils
Stirring utensil
Spatula
3 display plates
Display labels

Estimated Prep Time: 40 minutes per variation

G. Effect of Solid versus Liquid Fat on Quality of a Baked Product

Ingredients 3 c. flour, divided
1 t. salt, divided
1 c. sugar, divided
4 t. baking powder, divided
2 eggs, lightly beaten, divided
¼ c. vegetable oil
¼ c. shortening (do not melt)
1 c. milk, divided

2 c. fresh blueberries or 1 c. canned blueberries, divided

Equipment Oven with timer
Measuring utensils
Graduated liquid measuring cup
2 muffin pans
Mixing bowl per variation (2 variations)
Stirring utensils
Spatula
2 display plates
Display labels

Estimated Prep Time: 45 minutes per variation

H. Mayonnaise Variations

Ingredients 5 egg yolks, divided
7 T. vinegar or lemon juice, divided
1 ¾ t. salt, divided
¾ t. dry mustard, divided
¾ t. paprika, divided
1 ¾ t. sugar, divided
3 ½ c. vegetable oil, divided
½ c. melted butter
1 egg white
1 whole egg

Equipment Small bowl per variation (5 variations)
Electric mixer
Spatula
Graduated liquid measuring cup
Measuring spoons
Egg separator
7 display containers
1 freezer-safe container with lid
Freezer
Blender per variation (3 variations)
Display labels

Estimated Prep Time: 15-20 minutes per variation

I. Repairing a "Broken" Emulsion

Ingredients Egg white mayonnaise (from variation J.3)
1 egg yolk

Equipment Mixing bowl
Electric mixer
Spatula
Display label

Estimated Prep Time: 15 minutes

J. Evaluation of Oxidative Rancidity

Ingredients Rancid sunflower oil
Rancid sesame oil
Rancid peanut oil
Rancid vegetable oil
Rancid olive oil

Equipment Display labels

Estimated Prep Time: 10 minutes

K. Fritters

Ingredients 1 c. flour
1 ½ t. baking powder
½ t. salt
1 T. granulated sugar (use only in fruit fritters)
1 egg
½ c. milk
1 T. melted butter
½ c. of one of the following:

- drained, canned fruit (peaches, pears, fruit cocktail, crushed pineapple) **or**
- fresh, chopped fruit (apples, pears, bananas, peaches, blueberries) **or**
- drained, canned or cooked vegetables (corn, carrots, onions)

Oil for deep-fat frying

Equipment Deep-fat fryer
Sifter
Measuring utensils
Graduated liquid measuring cup
Large mixing bowl
Small bowl
Microwave (to melt butter)
Stirring utensil
Colander
Tongs
Paper towels
Baking sheet
Oven

Estimated Prep Time: 20 minutes

L-a. Easy Thousand Island Dressing

Ingredients 1 c. low fat mayonnaise
½ c. catsup
¼ - ⅓ c. sweet pickle relish

Equipment Container with a lid
Stirring utensil
Measuring utensils
Spatula

Estimated Prep Time: 15 minutes

L-b. Curried Sour Cream-Yogurt Dressing

Ingredients ½ c. low-fat cultured sour cream
½ c. plain yogurt
¾ t. curry powder
2 T. pineapple juice

Equipment Container with lid
Stirring utensil

Measuring utensils
Spatula

Estimated Prep Time: 15 minutes

L-c. Waldorf Salad with Cooked Dressing

Ingredients 3 unpeeled apples, cored and cubed
¼ c. raisins
1 c. celery, diced
½ c. chopped nuts of your choice
2 egg yolks
$^{2}/_{3}$ c. milk
3 T. fresh lemon juice or vinegar
1 T. butter
1 ½ T. sugar
¼ t. salt
2 T. flour
½ t. dry mustard
Water for double boiler

Equipment Cutting board
Knife
Egg separator
Salad bowl
Stirring utensil
Small bowl
Whisk
Double boiler
Range
Refrigerator

Estimated Prep Time: 30 minutes

Unit 16 – Cakes

A. Yellow Cake with Leavening Agent Variations

Ingredients 10 ½ c. sifted cake flour, divided
5 ¼ c. sugar, divided
1 ¾ c. shortening, divided
7 eggs, divided
3 ½ t. salt, divided
3 ½ c. milk, divided
3 ½ t. vanilla, divided
3¾ t. double acting baking powder, divided
2½ t. baking soda, divided
1 ¼ t. cream of tarter, divided
¼ t. cornstarch, divided
1 T. lemon juice, divided
Aerosol vegetable oil cooking spray
Flour for pan preparation

Equipment Sifter
Measuring utensils
Graduated liquid measuring cup
Spatula
3 mixing bowls per variation (7 variations)
Electric mixer
Toothpick
Cake knife
7 pH strips
7 cake pans (8" × 8")
7 display plates
Oven with timer
Display labels

Estimated Prep Time: 45 minutes per variation

B. Effect of Sweetener on Shortened Cakes

Ingredients 1 ¼ c. shortening, divided
1 ½ c. granulated sugar, divided
5 eggs, divided
5 c. cake flour (sift before measuring) , divided
5 t. baking powder, divided
1 ¼ t. salt, divided
1 2/3 c. milk, divided
2 ½ t. vanilla, divided
¼ c. 2 T. honey
Amount of artificial sweetener equivalent to ½ c. sugar (½ c. Splenda®)
Grease and flour (for pan preparation)

Equipment 2 mixing bowls per variation (5 variations)
Measuring utensils
Graduated liquid measuring cup
Electric mixer
Sifter
5 cake pans (8" × 8")
Oven with timer
Toothpicks
Cooling rack

Estimated Prep Time: 55 minutes per variation

C. Effect of Fat on Shortened Cakes

Ingredients ¾ c. shortening, divided
2 ½ c. granulated sugar, divided
5 eggs, divided
5 c. cake flour (sift before measuring), divided
5 t. baking powder, divided
1 ¼ t. salt, divided
1 2/3 c. milk, divided
2 ½ t. vanilla, divided
¼ c. butter
¼ c. oil
Grease and flour (for pan preparation)

Equipment 2 mixing bowls per variation (5 variations)

Measuring utensils
Graduated liquid measuring cup
Electric mixer
Sifter
5 cake pans (8” × 8”)
Oven with timer
Toothpicks
Cooling rack

Estimated Prep Time: 55 minutes per variation

D. Effect of Flour Type on Cake

Ingredients
4 ½ c. sugar, divided
1 ½ c. shortening, divided
6 eggs, divided
1 ½ c. cake flour
1 ½ c. all-purpose flour
1 ½ c. bread flour
1 ½ c. whole wheat flour
1 ½ c. rye flour
1 ½ c. buckwheat flour
2 T. 1 ½ t. double-acting baking powder, divided
1 T. salt, divided
3 c. milk, divided
3 T. vanilla
Aerosol vegetable oil cooking spray
Flour for pan preparation

Equipment
Sifter
Measuring utensils
Graduated liquid measuring cup
Spatula
3 bowls per variation (6 variations)
Electric mixer
6 baking pans (8” × 8”)
Toothpick
Oven with timer
Cake knife
6 display plates
Display labels

Estimated Prep Time: 45 minutes per variation

E-a. Basic Angel Food Cake

Ingredients
1 c. cake flour, sifted before measuring
1 ½ c. granulated sugar
¼ t. salt
1 ½ t. cream of tartar
1 t. vanilla
1 ½ c. room-temperature egg whites (8-10 large eggs)

Equipment
Sifter
Measuring utensils
Graduated liquid measuring cup
2 mixing bowls
Egg separator
Electric mixer
Spatula
Tube pan
Oven with timer
Bottle or another item to invert pan while cooling
Display label

Estimated Prep Time: 1.25 hours

E-b. Basic Sponge Cake

Ingredients
1 c. cake flour, sifted before measuring
1 c. granulated sugar
¼ t. salt
½ t. cream of tartar
5 large eggs, separated
3 T. cold water
½ t. vanilla or lemon extract

Equipment
Sifter
3 mixing bowls
Egg separator
Measuring utensils
Electric mixer
Spatula
Tube cake pan
Oven with timer
Bottle or another item to invert pan while cooling
Display label

Estimated Prep Time: 1.25 hours

E-c. Basic Chiffon Cake

Ingredients
2 ¼ c. cake flour, sifted before measuring
1 ½ c. granulated sugar, divided
3 t. baking powder
½ t. salt
¾ c. cold water
½ c. oil
5 egg yolks
1 t. vanilla or lemon extract
1 c. egg whites, room temperature
½ t. cream of tartar

Equipment
Sifter
3 Mixing bowls
Measuring utensils
Graduated liquid measuring cup
Electric mixer
Spatula
Tube cake pan
Oven with timer
Bottle or another item to invert pan while cooling
Display label

Estimated Prep Time: 1.5 hours

F-a. Grandma's Gingerbread

Ingredients
½ c. butter
½ c. granulated sugar
1 egg, well beaten
1 c. molasses
2 ½ c. flour, sifted
1 c. hot water
1 t. salt
1 t. ground cloves
1 t. ginger
1 t. cinnamon
1 ½ t. baking soda
Aerosol vegetable oil cooking spray

Equipment
Large mixing bowl
Electric mixer
Measuring utensils
Spatula
Sifter
Oven with timer
Toothpick
Cooling rack
9" square baking pan

Estimated Prep Time: 1 hour

F-b. Banana Nut Cake

Ingredients
1 stick margarine
1 ½ c. sugar
2 eggs
2 c. + 2 T. flour (divided)
2 t. soda
Pinch of salt
6 T. buttermilk
2 ripe bananas, mashed
1 c. chopped pecans
1 t. vanilla
Shortening
¾ c. shortening
¼ c. margarine
½ t. salt
1 t. cream of tarter
1 t. vanilla
¼ c. water
1 box powdered sugar

Equipment
Electric mixer
2 mixing bowls
Measuring utensils
3 9" round cake pans
Oven with timer
Cooling racks
Spatula
Cake plate

Estimated Prep Time: 1 hour

Unit 17 – Pastry

A. Effect of Type of Fat

Ingredients
2 ½ c. all-purpose flour, divided
1 ¼ t. salt, divided
2 T. refrigerated shortening
2 T. refrigerated oil
2 T. refrigerated butter
2 T. refrigerated margarine
2 T. refrigerated lard
7 ½ T. cold water, divided

Equipment
Sifter
Measuring utensils
Rolling pin
Mixing bowl per variation (5 variations)
Pastry blender or knife
Fork
Waxed paper
Knife
Baking sheet
Oven with timer
Ruler
Pastry guides
Bailey shortometer, if available

Estimated Prep Time: 30 minutes per variation

B. Effect of Type of Flour

Ingredients
½ c. all-purpose flour
½ c. pastry flour
½ c. whole-wheat flour
½ c. cake flour
1 t. salt, divided
8 T. refrigerated shortening, divided
6 T. cold water, divided

Equipment
Sifter
Rolling pin
Measuring utensils
Mixing bowl per variation (4 variations)
Pastry blender or knife
Fork
Waxed paper
Knife
Pastry guides
Baking sheet
Oven with timer
Ruler
Bailey shortometer, if available

Estimated Prep Time: 30 minutes per variation

C. Effect of Extent of Manipulation

Ingredients
1 ½ c. all-purpose flour
¾ t. salt
6 T. refrigerated shortening
4 ½ T. cold water

Equipment Sifter
Rolling pin
Measuring utensils
Mixing bowl per variation (3 variations)
Pastry blender or knife
Fork
Waxed paper
Knife
Pastry guides
Baking sheet
Oven with timer
Ruler
Bailey shortometer, if available

Estimated Prep Time: 30-35 minutes per variation

D. Preparation of Commercial Pie Pastries

Ingredients 1 frozen ready-to-bake pie crust
1 package stick-type pie crust
Ingredients needed to prepare stick-type crust (see instructions on package)
1 package dry pie crust mix
Ingredients needed to prepare dry pie crust (see instructions on package)

Equipment Knife
Oven with timer
Equipment needed to prepare stick-type crust (see instructions on package)
Equipment needed to prepare dry pie crust (see instructions on package)
Ruler
Bailey shortometer, if available

Estimated Prep Time: 20 minutes per variation

E-a. Turkey Pot Pie

Ingredients 2 $^{1}/_{3}$ c. flour, divided
1 t. salt, divided
1 c. Crisco®
6 T. water
$^{1}/_{3}$ c. margarine
$^{1}/_{3}$ c. celery
$^{1}/_{3}$ c. chopped onions
½ t. salt
¼ t. pepper
1 c. chicken broth
$^{2}/_{3}$ c. milk
3 c. shredded turkey or chicken
1 c. frozen mixed vegetables, thawed
2 c. potatoes, chopped

Equipment Sifter
Rolling pin
Measuring utensils
Graduated liquid measuring utensil
Mixing bowl
Refrigerator
Pastry blender or knife
Fork
Waxed paper
Saucepan
Range
Stirring utensil
Knife
Cutting board
Pastry guides, if available
9" pie pan
Oven with timer
Can opener
Potato peeler

Estimated Prep Time: 1.25 hours

E-b. Lemon Pie

Ingredients 1 c. flour
½ t. salt
¼ c shortening
3 T. water
3 eggs
1 c. + 2 T. sugar, divided
3 T. cornstarch (level)
¼ c. margarine, melted
¼ c. real lemon juice
1 c. whole milk
1 c. sour cream

Equipment Sifter
Rolling pin
Measuring utensils
Egg separator
Graduated liquid measuring utensil
2 mixing bowls
Electric mixer
Refrigerator
Pastry blender or knife
Fork
Waxed paper
Saucepan
Range
Stirring utensil
Pastry guides, if available
9" pie pan
Oven with timer
Can opener

Estimated Prep Time: 50 minutes to 1 hour

E-c. Pecan Pie

Ingredients 1 c. flour
½ t. salt
¾ c shortening
3 T. water
3 eggs
1 c. sugar

1 c. Karo syrup
1 c. pecan halves
¼ t. vanilla extract

Equipment Sifter
Rolling pin
Measuring utensils
Graduated liquid measuring utensil
2 mixing bowls
Whisk
Refrigerator
Pastry blender or knife
Fork
Waxed paper
Saucepan
Range
Stirring utensil
Pastry guides, if available
9" pie pan
Oven with timer

Estimated Prep Time: 50 minutes to 1 hour

Unit 18 – Candy

A-a. Thermometer Calibration

Ingredients 2 c. water

Equipment Candy thermometer(s)
Graduated liquid measuring cup
Saucepan
Range

Estimated Prep Time: 10 minutes

A-b. Cold Water Test for Sugar Syrup (Demonstration)

Ingredients 1 c. sugar
½ c. water
Ice water for cups

Equipment Candy thermometer
Measuring utensils
Graduated liquid measuring cup
Saucepan
Range
Spoon
Cup of ice water per variation (5 variations)
Display labels

Estimated Prep Time: 30 minutes

B. Preparation of Fondant

Ingredients 2 c. granulated sugar, divided
½ c. + 2 T. hot water, divided
1 T. corn syrup
Ice water (for cold water test)

Equipment Saucepan with lid per variation (2 variations)
Timer
Measuring utensils
Graduated liquid measuring cup
Candy thermometer per variation (2 variations)
Range
Wooden spoon
Cup of ice water
2 large wet platters
2 display plates
Display labels

Estimated Prep Time: 25-35 minutes depending on variation

C. Preparation of Fudge

Ingredients 2 c. granulated sugar, divided
1 c. whole milk, divided
1 T. light corn syrup, divided
2 oz. unsweetened chocolate, cut into pieces, divided
2 T. butter or margarine, divided
1 t. vanilla, divided
Butter (to prepare baking dishes)

Equipment 2 Small baking dishes
Measuring utensils
Graduated liquid measuring cup
Saucepan per variation (2 variations)
Candy thermometer per variation (2 variations)
Range
Wooden spoon
Cup of ice water
2 display plates
Display labels

Estimated Prep Time: 25-35 minutes depending on variation

D. Using Sugar Substitutes in Preparation of Noncrystalline Candy

Ingredients 1 c. granulated sugar
1 c. sugar substitute (Splenda®)
¾ c. corn syrup, divided
1 c. water, divided
1 t. lemon or mint extract, divided
Ice water (for cold water test)
Oil (for oiling cookie sheet)

Equipment Saucepan per variation (2 variations)
Candy thermometer per variation (2 variations)
Measuring utensils
Graduated liquid measuring cup
Range

Spoon
Clear glass
Cookie sheet
Spatula
2 display plates
2 display labels
Food coloring (optional)
Toothpicks (optional)

Estimated Prep Time: 30 minutes per variation

E. Evaluation of Nonnutritive Sweeteners

Ingredients Assortment of nonnutritive sweeteners (chosen by instructor)
Fruit (chosen by instructor)

*Suggestions: Equal®; Sweet'N Low®; Splenda®; PureVia®; Sunette

Equipment Tray
Small bowls for sugar substitute (based on number of samples)
Spoons (based on number of samples)
Tongs
Cutting board
Knife
Package label with ingredient listing and nutrient analysis
Price information for samples

Estimated Prep Time: Depends on number of samples used. Approximately 10 minutes for evaluation.

F-a. Peanut Brittle

Ingredients 1 ¼ c. sugar
½ c. corn syrup
½ c. water
2 T. butter or margarine
½ t. baking soda
1 c. unsalted peanuts
Ice water (for cold water test)

Equipment Saucepan
Candy thermometer
Spoon
Measuring utensils
Graduated liquid measuring cup
Range
Clear glass cup
Cookie sheet
Display plate
Display label

Estimated Prep Time: 30 minutes

F-b. Toffee

Ingredients 1 c. sugar
¾ c. + 2 T. butter
¼ c. water
2 T. corn syrup
¼ c. toasted, sliced almonds (optional)
Ice water (for cold water test)
Butter (for pan preparation)

Equipment Saucepan
Candy thermometer
Measuring utensils
Graduated liquid measuring cup
Range
Spoon
Clear glass cup
Baking dish or pie plate
Display plate
Display label

Estimated Prep Time: 20 minutes

G. Pralines

Ingredients 1 pkg. brown sugar
2 sticks margarine, melted
2 eggs
1 ½ c. flour
1 t. vanilla extract
1 c. pecan halves
Aerosol vegetable oil cooking spray

Equipment Oven with timer
Large mixing bowl
Electric mixer
Measuring utensils
9" square baking pan
Spatula
Microwave (to melt butter)
Microwave safe dish
Display plate
Display label

Estimated Prep Time: 45 minutes

H. Chocolate Covered Cherries

Ingredients 2 lbs. powered sugar
14 oz. condensed milk
½ c. margarine or butter
~3 doz. maraschino cherries
12 oz. chocolate chips
2 T. paraffin, grated

Equipment Large mixing bowl
Mixing utensil
Measuring utensils
Grater
Cookie sheet
Waxed paper
Foil
Colander
Toothpicks
Double boiler
Spoon

Refrigerator
Food handler gloves
Timer or clock
Range

Estimated Prep Time: 3 ½ - 4 hours including refrigeration time.

Unit 19 – Frozen Desserts

A. Vanilla Ice Cream

Ingredients 2 c. granulated sugar, divided
7 ½ c. whole milk, divided
5 c. whipping cream, divided
7 ½ t. vanilla, divided
5 pinches salt, divided
Sugar substitute equivalent to ½ c. sugar (Splenda®)
½ c. honey
1 envelope (7 g) unflavored gelatin
Crushed ice (for ice cream maker)
Rock salt (for ice cream maker)

Equipment Saucepan
Measuring utensils
Graduated liquid measuring cup
Stirring utensil
Range
Ice cream freezer per variation (5 variations)
Manufacturer directions
Freezer

Estimated Prep Time: Varies dependent on ice cream freezer.

B-a. Orange Sherbet

Ingredients 2 c. milk
½ c. frozen orange juice concentrate
½ c. sugar
½ t. finely grated orange rind
Pinch of salt
Crushed ice (for ice cream maker)
Rock salt (for ice cream maker)

Equipment Saucepan
Measuring utensils
Graduated liquid measuring cup
Stirring utensil
Range
Ice cream freezer
Manufacturer directions
Freezer

Estimated Prep Time: Varies dependent on ice cream freezer.

B-b. Low-Fat Vanilla Ice Cream

Ingredients 2 ½ c. 2% milk
½ c. granulated sugar
1 ½ t. vanilla
Pinch salt
Crushed ice (for ice cream maker)
Rock salt (for ice cream maker)

Equipment Saucepan
Measuring utensils
Graduated liquid measuring cup
Stirring utensil
Range
Ice cream freezer
Manufacturer directions
Freezer

Estimated Prep Time: Varies dependent on ice cream freezer.

C-a. Frozen Yogurt

Ingredients ½ envelope unflavored gelatin
½ T. lemon juice
2 T. cold water
1 c. strawberries, peaches, raspberries or cantaloupe
1 T. granulated sugar
1 c. vanilla yogurt

Equipment Double boiler
Measuring utensils
Range
Stirring utensil
Blender
Shallow metal pan
Foil
Freezer

Estimated Prep Time: 24 hours

C-b. Fruit Bavarian Cream

Ingredients 1 envelope unflavored gelatin
¼ c. cold water
¼ c. boiling water or juice from canned fruit
¾ c. sugar
¾ c. crushed fruit
¾ c. whipping cream

Equipment Saucepan
Range
Measuring utensils
Graduated liquid measuring cup
Stirring utensil
Spatula
2 mixing bowls
Stirring utensil
Electric mixer
Gelatin mold or metal pan
Refrigerator

Estimated Prep Time: 6 hours

Unit 20 – Beverages

A. Coffee Brewing Methods

Ingredients
- 2 T. drip-grind coffee*
- 4 c. water
- 4 T. regular-grind coffee
- 2 T. decaffeinated drip-grind coffee
- (*2 T. per 1 c. water = 1 serving)

Equipment
- Drip coffee maker with carafe
- 2 coffee filters
- Measuring spoons
- Graduated liquid measuring cup
- Saucepan
- Range
- Steeping container
- Cheesecloth
- Thermometer
- Percolator coffee pot
- 4 clear display coffee cups
- Display labels

Estimated Prep Time: 10-15 minutes per variation

B. Instant Coffees

Ingredients
- 1 serving regular instant coffee
- 1 serving regular freeze-dried coffee
- 1 serving decaffeinated instant coffee
- 1 serving decaffeinated freeze-dried coffee
- Water

Equipment
- 4 clear display coffee cups
- Saucepan
- Range
- Graduated liquid measuring cup
- Measuring spoons
- Spoon
- Display labels

Estimated Prep Time: 2-3 minutes per variation

C. Teas

Ingredients
- Boiling water
- 1 t. green tea leaves
- 1 t. oolong tea leaves
- 1 t. black tea leaves
- 1 t. decaffeinated tea leaves
- 1 t. herbal tea leaves

Equipment
- Teapot per variation (5 variations)
- Tea strainer
- 5 clear display coffee cups
- Saucepan
- Range
- Graduated liquid measuring cup
- Display labels
- Thermometer
- Timer

Estimated Prep Time: 10 minutes per variation

D. Effect of Acid on Tea

Ingredients
- 2 black tea bags
- 2 c. boiling water
- 1 T. lemon juice

Equipment
- Teapot
- Timer
- Saucepan
- Range
- Graduated liquid measuring cup
- Measuring spoons
- 4 custard cups
- Spoon
- Refrigerator
- Display labels

Estimated Prep Time: 60 minutes

E. Chocolate and Cocoa Beverages

Ingredients
- 2 T. cocoa
- 4 T. granulated sugar, divided
- 1 c. water, divided
- 3 c. milk, divided
- ½ t. vanilla , divided
- 2 dashes of salt, divided
- 2 envelopes commercial instant cocoa
- Ingredients needed to prepare cocoa according to package directions

Equipment
- Small saucepan per variation (2 variations)
- Measuring utensils
- Graduated liquid measuring cup
- Range
- Stirring utensil
- Thermometer
- Equipment needed to prepare cocoa according to package directions
- 3 clear display coffee cups
- Display labels

Estimated Prep Time: 15 minutes per variation

F-a. Viennese Coffee Mix

Ingredients
- ¼ c. hot cocoa mix
- ¼ c. nondairy creamer mix
- 3 T. instant coffee powder
- 2 T. powdered sugar
- $^{1}/_{8}$ t. cinnamon
- $^{1}/_{8}$ t. nutmeg
- 1 c. boiling water

Equipment Container with lid
Spoon
Measuring utensils
1 coffee cup
Graduated liquid measuring cup
Saucepan
Range

Estimated Prep Time: 5 minutes

F-b. Cafe Aruba

Ingredients 1 orange
½ c. coffee
1 T. sugar (divided)
1 qt. water
½ c. heavy cream

Equipment Vegetable peeler
Knife
Cutting board
Coffee maker
Measuring utensils
2 mixing bowls
Electric mixer
Spatula
Serving cups

Estimated Prep Time: 30 minutes

F-c. Hot Spiced Tea

Ingredients 1 qt. boiling water
¼ c. sugar
10 whole cloves
2 cinnamon sticks
4 tea bags
¼ c. orange juice
1 T. orange rind, grated
½ lemon, sliced thin and seeded

Equipment Saucepan
Range
Measuring utensils
Graduated liquid measuring cup
Stirring utensil
Timer
Tea strainer
Serving cups
Knife
Cutting board

Estimated Prep Time: 15 minutes

F-d. French Chocolate

Ingredients 1 ¼ square bitter chocolate
3 T. water
¼ c. + 2 T. sugar
½ c. heavy cream, whipped
1 qt. milk, scalded

Equipment Saucepan
Range
Measuring utensils
Graduated liquid measuring cup
Spatula
Stirring utensil
Serving cups

Estimated Prep Time: 20 minutes

F-e. Mexican Chocolate

Ingredients 2 c. milk
1 cinnamon stick
1 ½ square semisweet chocolate

Equipment Saucepan
Range
Graduated liquid measuring cup
Measuring utensils
Stirring utensil
Electric mixer
Serving cups

Estimated Prep Time: 20-25 minutes

F-f. Orange Julius

Ingredients 6 oz. frozen orange juice concentrate
½ c. sugar
1 c. water
1 c. milk
1 t. vanilla
Ice

Equipment Measuring utensils
Graduated liquid measuring cup
Blender
Stirring utensil
Serving cups

Estimated Prep Time: 10 minutes

Appendix C – Glossary

affective tests – Subjective tests that evaluate individual preferences. Examples include hedonic and personal preference tests.

aged cheese – Cheese made from the curd or solid portion of curdled milk that has <80% moisture and is held at a defined temperature and moisture for a designated time, with added curd turning or added ingredients to develop desired texture and flavors. Examples include cheddar, aged cheddar, Swiss, mozzarella, and provolone.

albumin – Egg white, which is comprised primarily of protein and water.

al dente – Refers to the texture of properly cooked pasta, which is tender, yet firm enough to resist the teeth.

analytical tests – Subjective tests that are used to detect differences. Examples include discriminative tests and descriptive tests.

astringent – Sour or bitter in taste, resulting in the puckering of mouth tissues.

au jus – Natural juices or gravy from meat.

Bailey shortometer – Instrument used to objectively evaluate the tenderness of pastry, crackers, or cookies by measuring the amount of force required to break a product.

bake – To cook in an oven. Covered or uncovered containers may be used. When applied to meats in uncovered containers, it is called *roasting*.

baste – To moisten foods while cooking, especially while roasting meat. Melted fat, meat drippings, stock, water, or fruit juices may be used.

beat – To make a mixture smooth or to introduce air by whipping, using a rapid, regular motion.

beurre manié – A soft paste thickener prepared by blending equal parts of butter and flour.

blanch – To dip food briefly in boiling water to inactivate enzymes or to remove skins from fruits, nuts, and vegetables.

blend – To thoroughly mix two or more ingredients together until they become one.

blind bake – Baking of an unfilled pie crust.

body – Describes the consistency of frozen desserts considering firmness, richness, viscosity, and resistance to melting.

boil – To cook in a liquid that bubbles actively during the time of cooking. The boiling temperature of water at sea level is 212 °F (100 °C).

braise – To brown meat or vegetables in a small amount of fat, then to simmer slowly in a small amount of liquid in a covered container. The liquid may be juices from meat, or additional water, milk, meat stock, or vegetable juices.

bran – Outer covering of the cereal grain that protects its endosperm.

bread – To coat foods by dipping into fine dry crumbs, into an egg-milk mixture, and then into crumbs again.

brew – To prepare a hot beverage by steeping or boiling.

broil – To cook food by direct heat under an intense heat source.

calibrate – To confirm or compare a measurement with an established standard.

candy – To cook in heavy syrup to preserve or glaze.

caramelize – To heat sugar, or foods containing sugar, until the sugar melts and a brown color and characteristic flavor develop.

casein – Major protein in milk that denatures and coagulates with exposure to an acidic environment or enzyme exposure.

chemethesis – The impression of being "hot" or "cold" given by certain foods that are not physically hot or cold when they are placed on the tongue. An example is hot peppers.

chiffon cake – Combination of a shortened and foam cake that includes fat from vegetable oil or egg yolk and egg foams. Cake flour and leavening are also ingredients in a chiffon cake.

chill – To place in a refrigerator or cool place until cold.

chop – To cut into small pieces with a knife or chopper.

clarify – To make or become clear or pure. Egg whites are often used to clarify liquids.

clarified butter – Butter that is treated to remove milk solids and water to reduce the potential for burning and increase the smoke point.

coagulation – The conversion of a liquid to a soft, semisolid mass.

coddle – To prepare eggs using a method in which an egg is cracked into a small cup called a coddler. The whole coddler is submerged in simmering water until the egg is cooked.

cold water test – Objective test used to determine a syrup's consistency. A small amount of syrup is placed in a cup of cold water and its softness or firmness is assessed.

cream – To beat one or more foods until pliable. Usually refers to the beating of fat and sugar with an electric mixer or by hand with a wooden spoon resulting in a creamy consistency and incorporation of air.

crumb – Interior cell structure of a baked product. Cell size, shape, and thickness are criteria for evaluation.

crystalline candy – Soft, smooth, creamy candy made of small sugar crystals. Examples of crystalline candy include fudge, fondant, pralines, and divinity. (Rock candy is an exception—it is hard rather than soft because it is made of very large sugar crystals.)

cube – To cut into approximately ¼" to ½" squares.

curd – The coagulated component of milk that results from the denaturation and coagulation of casein.

cut – To divide food materials with a knife or scissors. Common cutting styles include slicing, shredding, dicing, mincing, and peeling.

cut in – To distribute solid fat in dry ingredients with knives, with a pastry blender, or in a mixer with proper attachments.

dasher – Equipment placed within an ice cream freezer that stirs the ice cream mix while it freezes. The speed of the dasher determines how much air is incorporated into the mixture.

dextrinization – Process in which starch molecules are broken down into smaller, sweeter-tasting components due to exposure to dry heat.

dice – To cut into small, even-size cubes, usually ¼" or smaller.

dissolve – To mix a solid dry substance with a liquid until the solid substance is in solution.

double boiler – Cooking equipment that includes two pans with one pan inserted in the other. The food that is in the top pan is heated by boiling water in the bottom pan.

dredge – To sprinkle or coat with flour, sugar, or meal.

dry-heat preparation methods – Cooking techniques in which heat is transferred by air, radiation, fat, or metal. These include baking, roasting, broiling, grilling, barbequing, and frying.

endosperm – Largest component of the cereal grain that stores the starch and is the basis for flour production.

enzymatic browning – Browning that occurs when an enzyme acts on a phenolic compound in the presence of oxygen.

falafel – Spicy chickpeas that are ground, shaped into patties or croquets, and fried.

fermentation – The increase in size that the dough experiences as a result of carbon dioxide production and enzyme and pH alterations.

filet – A boneless piece of meat, fish, or poultry.

flake – To separate in small pieces. This is often used as an indicator of the doneness of fish.

foam cake – Cake prepared without or with a very small amount of fat from egg yolks; leavened with steam and air from beaten egg foam. Examples include angel food cake and sponge cake.

fold in – To gently incorporate one ingredient with another by hand. A large spoon or spatula is used to vertically cut through the mixture and turn it over by sliding the implement across the bottom of the mixing bowl.

fondant – Creamy, white crystalline candy that is used as icing to glaze and decorate various products and as the filling of many candies including chocolates.

fricassee – To cook by braising; usually applied to poultry or veal cut into pieces.

frizzle – To cook in a small amount of fat until crisp and curled at the edges, e.g., ham, bacon, dried beef.

fry – To cook in hot fat. When a small amount of fat is used, the process is known as *pan-frying* or *sautéing* or *shallow-fat frying*. When sufficient fat is used to keep food afloat, the process is known as *deep-fat frying*.

germ – Smallest, most nutrient-dense component of the cereal grain.

grill – To cook food by direct heat above an intense heat source.

grind – To reduce to particles by cutting, crushing, or grinding.

hummus – Paste or dip comprised of mashed chick peas, tahini, lemon juice, and garlic.

hydrogenation – Commercial process to make fats and oils more solid and extend their shelf life. In this process, hydrogen atoms are added to the double bonds in monounsaturated and polyunsaturated fatty acids to make them more saturated.

ice bath – Mixture of ice and water used to rapidly cool food or liquid products.

imitation cheese – Cheese-like product that contains vegetable oil instead of the milk fat found in natural cheese.

interfering agents – Ingredients added to a candy syrup that prevent sugar molecules from clustering together and forming large crystals.

julienne – To slice food, usually vegetables, into 1" to 3" sticks that are $^{1}/_{16}$" to $^{1}/_{8}$" thick.

kippered – Salted and smoked; refers to fish. Salmon and herring are often prepared using this technique.

knead – To manipulate—push, stretch, and fold—a dough to develop the gluten and create an elastic mass.

lamination – Alternating layers of fat and flour in pastry dough. Fat melts when the pastry is baked and leaves empty spaces where steam lifts the layers of flour and produces a flakey texture.

lard – To cover uncooked lean meat or fish with strips of fat to enhance flavor and reduce dryness during cooking.

leavening agents – Physical, biological, or chemical agents that cause a flour mixture to rise.

Maillard reaction – The reaction between a reducing sugar and protein resulting in the browning of foods such as baked products that develop a golden crust.

marinate – To cover with dressing or sauce and let stand for a length of time varying from less than an hour to overnight. Note: *Marinade* is the liquid or sauce in which the food is marinated.

mealy – Describes grainy pastry resulting from completely coating flour with oil.

mellow – Pleasing and full of flavor due to aging.

meniscus – The curved upper portion at the surface of a liquid in a container that is caused by surface tension. It is concave if the liquid is attracted to the container walls, but it is convex if it is not.

meringue – Egg white foam prepared by beating egg whites into foam and adding sugar. The quantity of sugar added determines the softness or hardness of the foam.

mince – To cut or chop fine using a knife or chopper.

mix – To combine two or more ingredients so that they are evenly distributed.

modified starch – A chemically or physically modified starch with increased functional characteristics.

moist-heat preparation methods – Cooking techniques in which liquids such as water, water-based liquids, or steam are used to heat foods and enhance their appearance and palatability. These techniques include scalding, poaching, simmering, stewing, braising, boiling, parboiling, blanching, and steaming.

mother sauces – Sauces that serve as a base from which other sauces are prepared. Groups of mother sauces include béchamel (white sauce), espagnole (brown sauce), hollandaise sauce, tomato sauce, and velouté sauce.

non-aged or **fresh cheese** – Highly perishable soft, whitish-colored, mild-tasting cheese that is made from the curd or solid portion of curdled milk, has >80% moisture content, and is not aged. Examples include cottage cheese, cream cheese, and ricotta cheese.

noncrystalline candy – Candy without organized structure or form; no crystals are present. Examples include caramel, peanut brittle, taffy, hard candy, and gummy candy.

oven spring – The quick expansion of dough during the first ten minutes of baking, which is due to expanding gases.

overrun – Increase in volume due to incorporation of air into a frozen dessert, resulting in a softer and creamier product.

pare / peel – To remove a skin or outside covering using a knife or mechanical peeler.

pastry guides – Tools used to control the thickness of a pie pastry when rolling it out.

percent sag test – Objective test used to measure gel strength. Higher percent sag represents a more tender gel.

poach – To cook in shallow, simmering hot liquid.

processed cheese – Cheese produced by combining different varieties of natural cheese and mixing them with other ingredients (emulsifiers).

proof – The rising or increase in volume of shaped dough that occurs through fermentation.

proof box – Equipment designed to maintain optimal temperatures and humidity for the fermentation and rising of a yeast dough.

puff pastry – Pastry with multiple alternating layers of fat and flour that puffs when baked. After baking, a puff pastry may be filled with various fillings.

rancid – Spoiled, and exhibiting the resulting off-tastes and odors; refers to fats or oils.

reconstitute – To restore a dehydrated product to liquid form through the addition of water. Nonfat dry milk is reconstituted for consumption.

rehydrate – To replace the moisture that was removed with drying.

retrogradation – Seepage of water out of a gel due to continual contraction of the bonds between amylase molecules. This is accelerated by freezing.

roasting – A dry-heat cooking method used primarily for meats and poultry. Roasted products are usually basted to maintain moisture and enhance flavor.

roux – A thickener prepared by cooking equal parts of fat and flour and used to thicken sauces, gumbo, soups, and gravy.

sauté – To fry using a very small amount of fat.

scald – To heat liquid, usually milk, until hot but not boiling.

scallop – To bake food, usually cut in pieces, with a milk-based sauce or other liquid.

score – To make shallow lengthwise or crosswise slits across the surface of food with a knife, fork, or other implement.

sear – To brown the surface of meat by a short application of intense heat to develop flavor and improve appearance.

shortened cake – Cake prepared with fat; usually leavened with baking soda or baking powder. Examples include pound cake, chocolate cake, butter cake, etc.

shred – To cut or tear into thin strips or pieces. Often a knife or a food processor shredder attachment is used to shred food.

sift – To put dry ingredients through a sieve.

simmer – To cook in liquid at a temperature just below the boiling point (185 °F to 210 °F).

slurry – A thickener prepared by blending starch and a cool liquid. This combination is then gradually mixed with a simmering liquid sauce base.

smoke point – The temperature at which fat or oil begins to smoke.

soft peak stage – Stage of foam formation at which peaks fall over as the beater is lifted from the foam.

sorbet – Frozen dessert prepared with pureed fruit or fruit flavoring and sugar syrup. Sorbet does not contain fat, eggs, gelatin, or dairy products.

soufflé – Modified omelet that includes a thick base such as a white sauce, egg white foam, and flavoring ingredients.

soybeans – Seeds from a plant in the legume family that is the rich protein source of tofu.

stabilizer – Ingredient added to food to minimize deterioration and loss of desirable properties.

steam – To cook in steam with or without pressure.

steep – To let stand in hot liquid below boiling temperature in order to extract flavor, color, or other qualities from a specific food.

stew – To simmer in a small amount of liquid.

stir – To mix ingredients using a spoon or electric mixer (on low speed) until dry ingredients are moist.

stir-fry – To fry by stirring and turning food very quickly and using a very small amount of fat.

stiff peak stage – Stage of foam formation at which just the tip of the peak bends over as the beater is lifted from the foam.

still-frozen dessert – Light, airy, smooth dessert that is not agitated during freezing and contains whipped egg whites or whipped cream, which prevents large ice crystal formation.

strain – To separate particles from liquid through use of a screen, sieve, colander, etc.

stroke – When stirring, one complete revolution around the bowl with the spoon (from 12 o'clock to 12 o'clock).

strudel – German pastry prepared with layers of thin dough filled with fruit or cream cheese. The pastry becomes very flakey when baked.

sunny side up egg – An egg cooked until the white is set and the yolk is still soft.

tahini – Thick paste comprised of ground sesame seeds.

threshold concentration – Concentration required to elicit a taste response.

toast – To brown the surface of food by application of direct or dry heat.

tofu – Cheese produced from soy milk.

toss – To mix lightly; usually refers to salad ingredients.

truss – To bind or fasten together before cooking to enhance the maintenance of shape while cooking. Examples include skewering meat or binding wings of poultry before cooking.

whey – Protein in milk that denatures in response to the application of heat. It is also the watery portion of milk removed from the curd in cheese production that consists of water, lactose, and whey protein.

whip – To beat rapidly to increase volume by incorporating air.

white sauce – Sauce comprised of fat, flour, milk or stock, and seasoning that is often served with vegetables, fish, or meat. When meat broth or fat from frying meat is used, it is called *gravy*.

Appendix D – List of Figures, Tables, and Graphs

Unit 1 – Sensory Evaluation

Unit 2 – Food Preparation Basics

Unit 3 – Meats

Unit 4 – Poultry

Unit 5 – Fish and Shellfish

- Table A-1 Fish Comparison

Unit 6 – Milk

- Figure 6-1: Funnel and graduated cylinder set up to measure syneresis.
- Table A-1 Sampling of Milk Products
- Table B-1 Sampling of Yogurt Products
- Table C-1 Evaluation of the Effect of Heat and Acid
- Table C-2 Evaluation of the Effect of Heat and Acid
- Table D-1 Effect of Tannins on Milk Proteins
- Table E-1 Comparison of Vanilla Pudding Products
- Table F-1 Milk Foams

Unit 7 – Cheese

- Table A-1 Evaluation of Cheese Products
- Table B-1 Effect of Heat on Cheese Products
- Table C-1 Evaluation of Cottage Cheese Curds
- Table D-1 Evaluation of Cheese Sauces

Unit 8 – Eggs

- Figure 8-1: Setting up a hot water bath.
- Figure 8-2: Beaten egg whites at the foamy stage.
- Figure 8-3: Beaten egg whites at the soft peak stage.
- Figure 8-4: Egg foams; soft peak stage on the left, stiff peak stage in the middle and on the right.
- Figure 8-5: Setting up an egg foam in a funnel for measurement of syneresis.
- Table A-1 Evaluation of Egg Quality
- Table B-1 Evaluation of Hard Cooked Eggs
- Table C-1 Cooking Scrambled Eggs
- Table D-1 Comparison of Baked and Stirred Custards
- Table E-1 Comparison of Egg White Foams
- Table F-1 Comparison of Angel Food Cake Products
- Table G-1 Evaluation of Brownies

Unit 9 – Vegetables and Fruits

- Figure 9-1: Correct use of a knife and cutting board.
- Table A-1 Enzymatic Oxidative Browning
- Table B-1 Comparison of Effects of Cooking Medium on Apples
- Table B-2 Comparison of Effects of Cooking Medium on Second Variety of Apples
- Table C-1 Effect of Cooking Treatment on Broccoli
- Table C-2 Effect of Cooking Treatment on Carrots
- Table C-3 Effect of Cooking Treatment on Red Cabbage
- Table C-4 Effect of Cooking Treatment on Cauliflower
- Table D-1 Factors Affecting Chlorophyll
- Table E-1 Effects of Apple Variety and Cooking Medium on Applesauce

Unit 10 – Legumes

- Table A-1 Evaluation of Prepared Legumes

Unit 11 – Cereals and Flour

- Table A-1 Comparison of Oatmeal Cereals
- Table B-1 Comparison of Types of Rice
- Table C-1 Comparison of Types of Pasta
- Table D-1 Comparison of Gluten Content

Unit 12 – Starches and Sauces

- Table A-1 Comparison of Vanilla Pudding Products
- Table B-1 Effect of Starch Variety on Lemon Pie Filling
- Table C-1 Effect of Type of Starch on Viscosity of Starch Pastes
- Table D-1 Effect of Acid and Sugar on Viscosity of Starch Pastes
- Table E-1 Evaluation of White Sauces

Unit 13 – Quick Breads

- Figure 13-1: Muffin batter, correctly stirred; note lumpy surface.
- Figure 13-2: Proper portioning of muffin batter.
- Figure 13-3: Using a ruler to measure muffin height, which relates directly to muffin volume.
- Figure 13-4: Over-stirred muffin batter—long gluten strands, smooth.
- Figure 13-5: Correctly stirred muffin on the left, and over-stirred on the right; exterior (top) and interior (bottom).
- Table A-1 Leavening of Muffins
- Table B-1 Leavening By Air
- Table C-1 Effect of Egg Protein on Popovers
- Table D-1 Effect of Manipulation and Flour Type on Muffins
- Table E-1 Effect of Manipulation on Biscuits
- Table F-1 Effect of Sugar on Cornbread

Unit 14 – Yeast Breads

- Table A-1 Preparation of Yeast Breads by Various Methods

Unit 15 – Fats and Oils

- Figure 15-1: Equipment setup used to measure a fat's smoke point.
- Table A-1 Comparison of Selected Commercial Fat Products Used as Spreads
- Table B-1 Effect of Coating on Fat Absorption
- Table C-1 Effect of Cooking Time on Fat Absorption at 375 °F
- Table C-2 Effect of Cooking Time on Fat Absorption at 350 °F
- Table D-1 Comparing Stir-Fry and Oven Fry Cooking Methods
- Table E-1 Polyunsaturated versus Monounsaturated Oils: Vegetable Oil and Olive Oil
- Table F-1 Comparing Fat and Fat Substitutes
- Table G-1 Effect of Solid versus Liquid Fat on Quality of a Baked Product
- Table H-1 Preparation of Mayonnaise Variations
- Table I-1 Repairing a "Broken" Emulsion
- Table J-1 Evaluation of Oxidative Rancidity

Unit 16 – Cakes

- Table A-1 Yellow Cake with Leavening Variations
- Table B-1 Effect of Sweeteners on Shortened Cakes
- Table C-1 Effect of Fats on Shortened Cakes
- Table D-1 Comparison of Flour Types
- Table E-1 Comparison of Foam Cakes

Unit 17 – Pastry

- Figure 17-1: Correct use of a pastry blender.
- Figure 17-2: Measuring a muffin's properties with a texture analyzer.
- Table A-1 Effect of Type of Fat
- Table B-1 Effect of Type of Flour
- Table C-1 Effect of Extent of Manipulation
- Table D-1 Comparison of Commercial Pie Pastry

Unit 18 – Candy
- Table A-1 Cold Water Test for Sugar Syrup
- Table A-2 Cold Water Test for Sugar Syrups
- Table B-1 Preparation of Fondant
- Table C-1 Preparation of Fudge
- Table D-1 Using Sugar Substitute in Preparation of Noncrystalline Candy
- Table E-1 Evaluation of Nonnutritive Sweeteners

Unit 19 – Frozen Desserts
- Table A-1 Homemade Vanilla Ice Cream
- Table B-1 Comparing Sherbet and Low-Fat Ice Cream

Unit 20 – Beverages
- Figure 20-1: A drip coffee pot.
- Figure 20-2: A tea pot for steeping tea bags.
- Table A-1 Coffee Brewing Methods
- Table B-1 Comparison of Instant Coffees
- Table C-1 Comparison of Teas
- Table D-1 Effect of Acid on Tea
- Table E-1 Comparison of Chocolate and Cocoa Beverages

Appendix A – Objective Tests and Templates
- Figure A-1: Place a 2" cylinder (biscuit cutter) on the linespread diagram and fill with substance.
- Figure A-2: Lift off cylinder and allow substance to spread for the designated flow period before taking the linespread readings using the numbered circles.
- Figure A-3: Insert a metal skewer or glass rod into the gel within its container.
- Figure A-4: Measure the depth of penetration into the gel using a ruler.
- Figure A-5: Invert the gel onto a flat surface and insert a skewer or rod to measure depth of penetration (either immediately or after allowing it to sag for a designated period).

Appendix E – Recipe Index

Breads, Quick and Yeast

Desserts

Candies

Beverages